FUNDAMENTALS OF TEACHING MATHEMATICS AT UNIVERSITY LEVEL

FUNDAMENTALS OF TEACHING MATHEMATICS AT UNIVERSITY LEVEL

B BAUMSLAG
Sweden

ICP
Imperial College Press

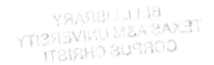

Published by

Imperial College Press
57 Shelton Street
Covent Garden
London WC2H 9HE

Distributed by

World Scientific Publishing Co. Pte. Ltd.
P O Box 128, Farrer Road, Singapore 912805
USA office: Suite 1B, 1060 Main Street, River Edge, NJ 07661
UK office: 57 Shelton Street, Covent Garden, London WC2H 9HE

British Library Cataloguing-in-Publication Data
A catalogue record for this book is available from the British Library.

FUNDAMENTALS OF TEACHING MATHEMATICS AT UNIVERSITY LEVEL

ISBN 1-86094-214-8

Printed in Singapore by Regal Press (S) Pte. Ltd.

In memory of my parents,
Braine and Kalman,
my first and best teachers

Preface

Mathematics is an important subject in the universities of the world, and it is therefore unfortunate that there are so few books and courses on the teaching of it at university level. As a larger percentage of the young are coming to study in the universities, teaching is becoming harder not easier. In my view this provides the justification for this book.

This book is concerned with the basics. Beginners and those who want an overview will benefit most. The book would also suit a variety of other readers: legislators and educational administrators, heads of departments, directors of undergraduate studies, even established lecturers. I feel this is natural because education cannot be duscussed piece-meal, instead the global picture must be kept in mind. Lecturers have considerable freedom, but are bound by decisions made by governments, states, their universities and departments. They are also strongly dependent on the interests, attainments, abilities and determination of the students, and thus can not afford to go ahead without keeping all these matters in mind.

The book is divided into four parts as follows:

- **PART I** Education in General: Chapters 1–4.
- **PART II** General Theory of Teaching: Chapters 5–7.
- **PART III** Departmental Matters: Chapters 8–10.
- **PART IV** The Individual Lecturer: Chapters 11–13.

Thus the topics are discussed down the chain of command, from the more powerful down to the less powerful, from society's over-riding demands, through the university and then the department, to finally the part that the lecturer can control.

This book can be used in a course in mathematics teaching or else for individual reading. Although I believe the order of the chapters is optimal, it is not necessary to read them in that order, since there are appropriate references. One can restrict one's reading to Chapters 6–7, and 11–13 for practical steps to improve teaching.

I believe that academics must take a much greater interest in education as a whole, because they are well suited to help formulate education policies, and that on the whole they have not taken this responsibility seriously. I think it is sensible to have a global view of the education system in one's own country, and to have some idea of at least one other country. Thus some examples of how this can be done are given in Chapter 1 and Appendix A.

The rise in both the number of universities and the number studying at university throughout the western world is striking. This poses opportunities, but also brings dangers. It is possible that the best students will be neglected due to the obvious importance of helping the weaker ones. And obviously there should be a greater variety of courses to match the greater variety of student ability, attainments and interests. But there should always be deep and demanding courses for the best students, otherwise mathematics will gradually die.

Since in many countries schools attended before university have ceased to regard one of their aims as matching their final courses to the university's requirements, I feel it is important for lecturers to obtain as clear a knowledge of the final years of school mathematics as possible.

Teaching in my opinion will be improved if the university and the department take a greater interest in organisation and how it can be used to improve teaching.

We mathematicians should make an effort to formulate and teach study skills, and I have several suggestions. There is a series of steps for studying a book (briefly described as SQERPSR^2), a technique for linking ideas together (City Explorer), and what I call the Jigsaw Puzzle method, which aids progress even when the overall picture is confused. I emphasise the value of a summary book, the value of tables and diagrams, and the role of memory and mnemonics. Students should be encouraged not only to learn a theorem but also to learn an example to illustrate it; the two should always appear together in the student's mind.

As a practical means of analysing teaching I propose ten fundamental rules of teaching, and also suggest a systematic but simple system of recording results from one year to the other.

Elegant proofs and arguments are all very well, but even more important are the ways used to help construct these ideas, which often are omitted from the final version. There are some useful techniques for lecturing, like writing an example on one side of the blackboard, and the general proof on the other side, and matching them step for step as far as possible.

Comprehensive and intelligible previews are important. Thus for instance I introduce point set topology with a fable of the mathematician who had his ruler stolen while out to lunch. As the course progresses and the student learns more ideas and concepts, a more precise and detailed overview should be given.

There are new methods of teaching now coming into use, but I remain convinced that much can be done by simple means, that attention to even small details can make a useful difference. Unfortunately there are many small details, and it is hard to keep them all in mind simultaneously. Having them listed is a useful first step.

I believe one should spend considerable time trying to deal with topics that one knows from one year to the next will turn out to be tricky. There is for instance a method of making the epsilon delta definition of limit more concrete, continuity can be illuminated by betting on the horses, and an analogy with spices can help convey the idea of linear independence.

Our object as mathematicians is deep and genuine knowledge. Analogies and rough ideas should appear often and freely, but one must ensure the students know what is an intuitive explanation and what is a correct and precise mathematical formulation. This can be done for instance with marking the rough ideas by a special symbol.

It is important to relate the teaching to the students' interests, so at least mentioning the applications which appeal to the groups being taught is helpful, and likewise reference to the history of mathematics, and striking quotations from the great mathematicians of the past do help to increase motivation. On the whole simple techniques and methods which do not require special equipment are best.

Thus this book is a practical how-to-do-it book, and also a theoretical book. It raises issues which need to be discussed and gives a suitable framework for this discussion.

My comments will all basically be about mathematics at university. I am talking about my own experiences and thoughts from1960 till 1998.

Most of my lecturing career was spent at Imperial College, London, where I was a lecturer for 25 years. I had begun my teaching career in the University of Witwatersrand in South Africa, where I had been a student. I then proceeded as a Ph.D. student under Professors Hirsch and Gruenberg at Queen Mary College, London. After my Ph.D. I taught for a year at the University of South Wales at Cardiff. I then spent two years at the Courant Institute in New York, returning to Imperial College. After leaving Imperial

College I spent six months at the University of Lund, and since then five years at Mälardalen University, one of Sweden's newest universities. These represent a large variety of different types of universities, but even so, because of the large variation in universities, it is not possible to say something definitive for all universities. Yet much of what I say would have some application to most.

My purpose in this book is to state, underline and remind all concerned with the teaching of mathematics, of the fundamental measures which are needed. Many of the remarks I make may strike some readers as being obvious or just plain common sense[1], but in practice it is difficult for teachers to effectively achieve even those things that most would agree are obviously sensible. There must always be a great deal of compromise.

There are many mathematicians who put an enormous amount of energy and passion into their teaching and teaching methods, and it has not been possible to present their ideas in this book. To do so, even briefly, would have required a much larger book than this one. I hope that the bibliography at any rate will help the reader explore some of those ideas.

Contents

Preface vii

Acknowledgements xvii

PART I EDUCATION IN GENERAL

1. **Education Systems in Brief** 3
   - 1.1 An education system in brief 4
   - 1.2 What university teachers expect 6
   - 1.3 Checklist 7
   - 1.4 Summary 8

2. **The Expansion of Education** 11
   - 2.1 Education acts 11
   - 2.2 The step to mass education 12
   - 2.3 More students means worse students 14
   - 2.4 More students means different students 16
   - 2.5 Attainment levels 17
   - 2.6 Standards 19
   - 2.7 What about the smart students? 20
   - 2.8 Conclusions 21

3. **Aims** 23
   - 3.1 Education and employment 23
   - 3.2 Aims of society in supporting the university 26
   - 3.3 Aims of students, educationalists and academics 28
   - 3.4 Why learn mathematics? 32
   - 3.5 Conclusions 35

4. Universities and Government 37
4.1 Government requirements 37
4.2 The lecturers 39
4.3 Summary 43

PART II GENERAL THEORY OF TEACHING

5. Teaching 47
5.1 Improving teaching 47
5.2 Content 52
5.3 The McLone report 57
5.4 Modern technological methods of teaching 59
5.5 Calculators, computers and lists of formulae 62
5.6 Conclusions 64

6. Study Skills 67
6.1 Steady study 67
6.2 Memorising is important 68
6.3 Reading a book 69
6.4 Solving problems 73
6.5 Polya's ideas 74
6.6 The student's most common fault 75
6.7 The problem of definitions 76
6.8 How to use lectures and tutorials 77
6.9 Techniques of final revision 79
6.10 Summary 79

7. Rules of Teaching 81
7.1 Theories of learning 81
7.2 The fundamental rules of teaching 83
7.3 Seminars 91
7.4 Some extra remarks 93

PART III DEPARTMENTAL MATTERS

8. Organisation and Examinations 97
8.1 Organisation 97
8.2 Examinations 99
8.3 What exactly is your examination supposed to test? 104
8.4 A much neglected teaching aid 104
8.5 Directed teaching (also called spoon-feeding) 105
8.6 External examiners 108
8.7 Summary chapters 1—8 108

9. Planning 111
9.1 Fundamentals of departmental organisation 111
9.2 Mentors for students and staff 113
9.3 Department leadership 114
9.4 Timetables 116
9.5 Study skills 118
9.6 Deciding which are to be methods courses 118
9.7 Summary 118

10. Methods of Teaching and Equipment 119
10.1 How to introduce the axiomatic method 119
10.2 Techniques of study 120
10.3 The role of books 121
10.4 Using computers in the teaching of mathematics 124
10.5 The overhead projector and the blackboard 126
10.6 Different methods of teaching 127
10.7 Indirect methods of teaching 134
10.8 Conclusions 135

PART IV THE INDIVIDUAL LECTURER

11. Lecturer's Approach 139
11.1 Introductory remarks 139
11.2 The lecturer's approach 140
11.3 Put in the construction lines 143
11.4 Introduce, summarise, recapitulate, revise 143
11.5 Ask questions 149

11.6 Motivate ... 150
11.7 Introducing new concepts ... 152
11.8 Informal and formal definitions ... 154
11.9 The Good Picture and the Essential Example ... 155
11.10 Repetition and maxims ... 158
11.11 Degree of generality ... 159
11.12 Organising breaks in a lecture ... 160
11.13 Tutorials ... 161
11.14 Various types of tutorial ... 162
11.15 A price to pay ... 164
11.16 Summary ... 164

12. Some Practical Points ... 165
12.1 Your voice ... 165
12.2 Notes ... 166
12.3 How to discuss the statement of a theorem ... 168
12.4 Notation ... 171
12.5 Equivalence relations and functions ... 172
12.6 Limits ... 174
12.7 Vector spaces etc. ... 176
12.8 Guess-check ... 180
12.9 The teacher is wrong ... 181
12.10 Summary ... 181

13. Assessment of Teaching ... 183
13.1 The need for external control ... 183
13.2 Records ... 185
13.3 Assessing new methods of teaching ... 187
13.4 Assessing as it is done now ... 188
13.5 Fundamental indices ... 190
13.6 Balance ... 190
13.7 Rules when assessing a new method of teaching ... 191
13.8 Comparing the quantity of material in courses ... 192
13.9 Assessing the difficulty of examinations ... 194
13.10 Southampton University's approach ... 195
13.11 Summary ... 195
13.12 Epilogue ... 196

APPENDICES

Appendix A Education Systems in Brief 199
A.1 The English education system 199
A.2 The Swedish education system in brief 200
A.3 The German education system 202
A.4 The French system in brief 204

Appendix B Videos 207
B.1 Videos 207
B.2 From the AMS 207
B.3 From the MAA 211
B.4 From the LMS 212

Appendix C Quotations 213
C.1 Quotations 213

Appendix D Quick Questionnaire 219
D.1 Quick questionnaire 219

Appendix E Alternative Courses 221
E.1 Alternative courses 221

Appendix F Cartoons 223
F.1 Cartoons 223

Appendix G Maxims 225
G.1 Time, organisation and study 225
G.2 What is important 225
G.3 How to study 226
G.4 Solving problems 226
G.5 General advice to lecturers 226

Appendix H Projects 229

Bibliography 233

Index 237

Acknowledgements

A large number of people have helped in the writing of this book. In some cases they will be horrified to see that instead of following their advice, I have re-iterated my previous points of view. Their remarks however have not been ignored. As a result of them I have been forced to re-consider my views and to judge whether I could justify them. In some cases they have made me change my approach and in other cases they have led to new ideas. Often they have given me extra encouragement. In short, they have helped me to think, and I am very grateful to:

Ulla Axner, Piotr Badziag, David Baumslag, Pia Baumslag, Peter Bassarab-Horwarth, Richard Bonner, Lennart Egnesund, Trevor Hawkes, Geoffrey Howson, Milagros Izquierdo, Gareth Jones, Stefan Johansson, Indy Lagu, Lars-Göran Larsson, Frank Levin, Sven Levin, John Navas, Tim Porter, Oliver Pretzel, and Mikael Stenlund. In addition, I thank the referees for their suggestions and comments.

A special thanks to the head and the staff of the Department of Mathematics at the University of Calgary, who at short notice, and with great friendliness and generosity, gave me both urgently needed moral support and office and computing facilities which enabled me to begin this venture.

Part I Education in General

Chapter 1 Education Systems in Brief

Chapter 2 The Expansion of Education

Chapter 3 Aims

Chapter 4 Universities and Government

1. Education Systems in Brief

Preview

One of the theses of this book is that one should look at teaching globally, that it does not suffice to consider only for instance the individual teacher or lecturer.[2] I therefore recommend looking at the overlying system.

I feel it is important when teaching a subject to know how it fits into the general scheme of education, because education is cumulative, and higher education depends strongly on the education at school, and higher education leads on to other things in life. Consequently one needs to have an overall understanding of the educational system. I do not think it is sufficient to concentrate on the educational system in one's own country alone. Looking at several educational systems gives one something to compare with.

If you are a lecturer from another country or even another part of the country, or if you have graduated from school some time ago, your own recollections of schooling may be inappropriate or out of date.

Many academics take the view that there is no point in looking at the education system in their country; after all it is given, and there is nothing much that can be done with it. Yet if the education system is to improve, somebody must take the responsibility of considering the system in general, and if academics do not do that, who will? The answer is a very small group of educators and state or government officials, and they alone may not get the right approach.

Then too it is important to know the specific mathematical topics that students have studied at school, because this obviously influences the way you begin your first year lectures. In the past many schools arranged their courses to suit the requirements of the university. But nowadays arranging the school course to fit in with the university is no longer regarded as a fundamental aim of school education in many countries. Consequently it is

[2] I use these terms as synonyms, even though teaching may involve activities other than lecturing.

important for universities to check what the incoming students have studied at school, and take measures accordingly. This is the topic of the final sections of this chapter.

1.1 An education system in brief

I think it is important for each lecturer to carry out a brief summary of the education system as it affects his own university. It is also interesting to study the methods of other countries. Sometimes they differ quite radically, and there may be interesting points worth noting

In this section I give, as an illustration of what I mean, a very brief account of the education system in the USA. Because of the books and other materials that the Americans supply throughout the world the American system has a considerable influence on teaching everywhere. Some other countries are discussed in Appendix A.

The important thing to realise about the United States is the great diversity of both the schools and the universities. It is difficult to say something that holds in general.

Schools vary in different States and school districts. Attendance at school is compulsory up to age 16. Although there are some exceptions, schools do not achieve a particularly high academic standard compared with Europe. American universities will often start with pre-calculus courses, more or less schoolwork, which the university suspects has not been learnt satisfactorily.

Again, it is difficult to say something which is valid for all universities. One must bear in mind that each State is independent and could pursue different policies, and in any case universities can be independent of the State. There are something like 3½ thousand institutions, and something like 14 million students enrol each year. The number of bachelor degrees awarded annually is over a million.

The size of some American universities gives pause for thought. For instance, Ohio State has nearly 100,000 students. Even their football stadium holds 85,000 people. (Bear in mind also that football is an important aspect of an American university; this shows how cultural differences can play a vital role in a university. The football team is also an important source of income for many universities.) Standards of universities vary enormously. The United States has some universities with the highest standards and academic achievements in the world. It also has a number of mediocre universities.

Perhaps one fundamental distinction is between universities and two-year junior colleges, which correspond more to say the English A-levels, rather than an English university.

Another way of distinguishing between them is the selection procedures.

The first group of universities have an open door policy, that is, they accept everybody, and offer remedial courses. However, the drop out rate is high.

The majority of universities require graduation from high school, and 12 units of preparatory work. Entrance tests are not required. Drop out rates tend to be high, perhaps as much as 70%.

A third group of universities, numbering perhaps 200-300, require high school graduation and 16 units of preparatory work, and take account of the grades obtained when selecting. Drop out rates are about 20%.

The final group of universities select on a competitive basis. There are perhaps 100 such universities. Drop out rates are low.

Admission can be determined in a number of different ways. A fairly common method is the standardised test SAT. At other universities there is an admission examination, still others may base their admission on the result of school performance. Rutgers for instance admitted only the top 15% from the various schools.

A further complication in admission policies is that universities must guard against being accused of discriminating against say American Africans, or Hispanics, or women. For instance, some claim that the SAT is biased towards white American students.

A summary of the average student schooling is given in Fig. 1.1. Thus the first degree, a bachelor's degree, usually takes 4 years, with engineering possibly taking 5 years. The masters would take another one or two years; some universities require dissertations as well as course work, but normally course work is all that is needed. The doctorate normally takes an extra two or three years, so that one gets one's Ph.D at about 27 or 28.

Students are financed by parents and occasional scholarships, and most need to work as well. Usually students need to pay tuition fees, and an education can be very expensive, e.g. at Harvard tuition and living expenses are about \$15,000-\$20,000 per year (1996 estimate). Many of the state universities have relatively low or no tuition fees. Graduate students often get extra money through grants.

Age	Type	Qualification
6-12	Primary school	None
12-15	Junior high school	None
15-18	Senior High School or High School.	High School diploma
18-20	Junior college	
18-22	University or four year junior college	Bachelor's degree
23-24	Masters degree	Master's degree
23-27	Doctorate	Ph.D.

Fig. 1.1 Summary of the American Education system

1.2 What university teachers expect

As already mentioned, the idea that the final year of school should fit the students for the first year of mathematics is no longer automatic. The sort of detailed, intricate calculations involving trigonometric functions and logarithms, exponents, factorising and manipulation of polynomials that lecturers imagine are quite standard, have probably never been studied by many of the average students. Then in order to manage a pass at school, students may need only half the material, and so, taking into account that much fades from the memory, we are led to conclude that the students know (probably imperfectly) only a quarter of what we expect them to know perfectly.

The vast majority of university lecturers do not check the school syllabus from time to time to see what sort of material the incoming students have studied. This means that they can not be sure that they have carried out what I call the first fundamental rule of teaching (§7.2), namely that the teaching must begin at an appropriate level.

Lecturers tend for the main part to rely on their own recollections of what they learnt when they were at school. But for many of the reasons I shall give later (see Chapter 2), the situation is very different now. So I suggest that some time and effort be spent on checking what the school syllabus is at the present time. Looking at the syllabus and the textbooks can be very revealing, and need not take a long time.

Another method of understanding the level of achievement of the incoming students is to discuss them with a teacher who is currently teaching the final years at school. Armed with a list of topics and typical questions check which topics are included and at which level.

Of course one does need to check the school syllabus every year, but it would be sensible if the members of staff did so in turn, writing a brief report. Thus they might only need to do the job once every five years, with yearly up-dates from their colleagues.

Until you and your colleagues have had a chance to look at the school syllabus yourselves, it may be helpful to consider the following section.

1.3 Checklist

The following[3] used as a checklist will enable one to gauge the knowledge of the students coming from school. It is likely that you will find many of the topics, most of which we take for granted, are unknown to your students.

- ALGEBRA

(1) Algebraic manipulation, factorisation, $x^n—y^n$.
(2) Completing the square. Polynomial division. Inequalities. Absolute value.
(3) Rational functions.
(4) Partial fractions.
(5) $(1+x)^n$ for small x.
(6) Series.
(7) Permutations and combinations.

- GRAPHS

(8) Drawing graphs, straight line, parabola, ellipse, hyperbola, circle etc.
(9) Translation of axes to parallel axes.

- TRIG AND LOG FUNCTIONS

(10) Rules of exponents, e and logarithms.
(11) Trigonometry, Radians, three or six trigonometric functions, $\sin(x + y)$ etc.

---

[3]A modification of a list in LMS [1995]

(12) Sine and cosine rules.
(13) $a\cos x + b\sin x$ expressed as $r\cos(x + c)$.
(14) General solution of trigonometric equations.
(15) Small trigonometric approximations.
(16) Inverse trigonometric functions.

• CALCULUS
(17) Derivatives, rules of differentiation, derivative of the composition of two functions.
(18) Standard integrals, change of variable, integration by parts.
(19) Implicit differentiation.
(20) Parametric differentiation.
(21) Normals.
(22) Small increments.
(23) $\arctan(x)$ and $\arcsin(x)$ expressed as integrals.
(24) Volumes of revolution.
(25) Differential equations.
(26) Newton-Raphson approximation.

• VECTORS
(27) Vectors.
(28) Scalar products.
(29) Vector equation of a line.

• MISCELLANEOUS
(30) Complex numbers.
(31) Probability and statistics.

The list is a useful one to keep in mind. How many of these ideas do you automatically assume your students know? A little checking will enable you to have a more informed idea of your students' pre-knowledge.

1.4 Summary

An overall understanding of the education system, of which lecturers form an integral part, is necessary if education is to be successful. There may have been many changes since lecturers left school, and they run the risk of being out of date with current school practice.

Transitions from one system to another are always difficult, and therefore special care is needed at the junction from school to university.

Thus I urge lecturers to take the time and effort to understand the whole picture. Finally I refer the reader to the Third International Mathematics and Science Study for a fascinating comparison of school mathematics and science in various countries, and Howson [1998] for a discussion of such comparative studies.

2. The Expansion of Education

Preview

The most striking fact about education is that it has expanded so rapidly. Student numbers have increased, the age of staying on at school has increased, the numbers going on to university have increased as well, while the number of universities has grown, and the variety of degrees and courses has also increased. All this has occurred very quickly. University staff and traditions are slow to change, and in many ways they have not adjusted. In particular, a much more varied intake of students has not resulted in a much larger and varied choice of mathematics courses.

Previously competition for university places was extremely high. Now that so many more students are being admitted, it is inevitable that the pass standard has dropped. More students are being educated, albeit to a lower standard. An advantage is counter-balanced by a disadvantage. It is a matter of judgement whether we are better off on the whole. Democratisation of the university means that more people have an equal opportunity to compete for the most rewarding jobs, an important political and social aim for many governments and states. But there is also a danger, the danger that the best students will be swamped among the mediocre students and not be encouraged to reach their potential.

2.1 Education acts

It is worth while turning our attention to schooling for a short while. I restrict my attention to England. If you are not teaching in England, then I hope you will make yourself familiar with your own country's education history, perhaps using this chapter as a suitable model. England of course has had an enormous influence on the education systems of a variety of countries, but it is included mainly as an illustration.

If one traces the successive Education Acts one sees that

(1) Compulsory education was introduced in 1870, becoming effective in 1880.
(2) This was extended to 16 in 1947.
(3) School after the age of 16 is not compulsory, but the number staying on till 18 is increasing; in 1996 it was 37%.

It seems likely that the longer schooling is required because the increasing sophistication of business and industry requires educated people. Schools adjust to this need by providing the training which it is felt will be of help for business and industry.

2.2 The step to mass education

The universities have a long and complex history, and are very varied.

A short, interesting, and readable account is available in Brittanica [1989], beginning with the Salermo medical school in the 9th Century, and the first real university at Bologna in the 11th Century. The next important institution to arise was the University of Paris at 1150-1170. The University of Oxford appeared at the end of the 12th Century, with a group of dissatisfied students proceeding to Cambridge in 1209.

In the 1790's most universities had a core curriculum consisting of Grammar, Logic, Rhetoric, Geometry, Arithmetic, Astronomy and Music. Students would subsequently study medicine, law, or theology. The final examinations tended to be gruelling, and many students failed.

The first modern university was Halle in 1694, which renounced religious orthodoxy and reasoned by rational and objective intellectual enquiry, with lectures being held in German and not Latin as had been the case before.

I will not attempt to summarise and explain the development of the modern university, but content myself by pointing out that on the whole, admission to European universities has been for the few. Nowadays however, the university is going towards educating more and more of the population. For instance, in England the student population in 1996 was nearly 2½ times the student population in 1984[4], so that in 1996 there were 730,000 students in a population of 52 million, i.e. about 1.4% of the population were university students.

In the United States the enrolment figures were 3.6 million in 1959 and some 8 million in 1969. Thus the United States explosion in higher education occurred some 25 years earlier than the English explosion. Since 1969 higher education in the United States has grown steadily to some 14.3 million in 1995, i.e. the student population was nearly 1.8 times the student population in 1969. The United States population was approximately 250 million in 1995, so that nearly 6% of the population were students.

An interesting account of the developments in education in Britain is provided by Halsey [1995]. Among other things he points out that the age at which selection was carried out has constantly been rising. Up till some thirty years ago, a child's educational route was determined at the age of 11, the notorious 11+ examination deciding which type of secondary school one was assigned to.

As Halsey remarks (p. 175), the number of home students in higher education divided by the number of 18 year olds in the population was 7.2 per cent in 1962 (a crucial date when the important Robbins report on higher education was published), but rose to 20.3 per cent in 1990. It is projected to increase to 32.1 per cent in the year 2000. (I have blurred some distinctions made by Halsey.)

Halsey suggests that higher education is beginning more and more to mean simply any education after compulsory education has been completed, much as it does in America.

As Halsey points out (p. 174, 175), there are different ways of viewing this: "Thus for the traditionalist, the experience of higher education has been one of dilution, presented as expansion. For the reformer, on the contrary, access has widened to offer opportunities appropriate for a modern economy and an increasingly well-educated citizenry. For the traditionalist more looks worse. For the reformer more means different." Perhaps both the traditionalist and the reformer have some merits in their views, and we shall discuss both viewpoints in the two following two sections.

---

[4] 730,000 compared with 292,000.

2.3 More students means worse students

To begin with it we look at the situation from the viewpoint of the traditionalist. From this viewpoint, students who come into the universities nowadays from school are not well prepared.

It has been argued that in the past the academics have dictated the school syllabus. Thus their suggestions for a suitable school syllabus began with a "top-down" approach. First one decided what sort of mathematics one needed to study in the final year of the university. The mathematics for the preceding year should leave the student with the prerequisites for the final year, and so on, till one decided what was needed for entry to the university. It was this that would determine the last year at school, and all the other years would be arranged accordingly.

This approach to the study of mathematics has been rejected by many, since it was felt unreasonable that the studies of the majority should be determined by reference to the needs of the few who would subsequently study mathematics at university.

Schools based on this top-down approach had strict and demanding courses, which a large proportion of students found unpalatable and failed. Increasing efforts to get students to pass met with little success. Consequently these courses were modified. The top-down approach was rejected.

For instance, in Higginson [1988], we read:

"The most fundamental error in the traditional GCE A-level system was that each stage was designed to be suited for those who were going on to the next. School children who were not good enough to go on were regarded as expendable. The other view, which seems to be held in every other advanced country, is that each stage of education should be designed for the main body of those who take it and the following stage has to start from where the previous ended."

In other words, the school system must now be designed so that the average and below average student can cope with it. The student talented in mathematics has, so as to speak, to tread water in all his school years.

The effect of this is that it takes the talented students longer to learn a reasonable amount of material, reduces their ambition, and the students who have completed their school mathematics do not have those prerequisites needed for their university courses.

As Higginson says, in the past only the good students were catered for at school. Now the good students find their courses easy to pass, even boring, and never ever learn to study hard.

In the report, "Tackling the mathematics problem," LMS [1995] the authors make the following remarks:

"There is unprecedented concern among academics about the decline in the mathematical preparedness of those entering undergraduate courses in science and engineering. This can be seen, for example, in the reports of the Engineering Council (1995), the Institute of Physics (1994), and the Institute of Mathematics and its Applications (1995).

Mathematics, science, and engineering departments appear unanimous in their perception of a qualitative change."

The changes at school level in the philosophy of teaching, sensible as they are to prevent unnecessary damage to young personalities and to provide equality of opportunity, have caused considerable changes to student attitudes. In practice, we find that students who have just got the minimum pass at school can not cope with our courses, indeed, they are nowhere near to coping.

I am often reminded, when marking some unfortunate and desperate student's examination script, of W.W. Sawyer's striking remark (Sawyer [1943]):

"It would, I suppose, be quite possible to teach a deaf and dumb child to play the piano. When it played a wrong note, it would see the frown of its teacher, and try again. But it would obviously have no idea of what it was doing, or why anyone should devote hours to such an extraordinary exercise. It would have learnt an imitation of music. And it would have learnt to fear the piano exactly as most students fear what is supposed to be mathematics."

Ramsden [1992] puts the position very clearly: "Today's lecturers are expected to deal with an unprecedentedly broad spectrum of student ability and background. Detailed previous knowledge, especially in mathematics and science, cannot any longer be relied on. As a result, courses and teaching methods must be amended to deal with classes that are now not only larger, but also more mixed in their attainments."

2.4 More students means different students

The consequence of the changes in schools, the changes in syllabus and what is expected of the students there, the reduced entrance requirements, the need to admit more students, has led to the admission of weaker and less well prepared students.

The universities of the United States have been subject to pressures to admit more students for a much longer period than most of the European universities. These problems of democratisation in the United States were already present in the 60's and provoked resentment. "Yet our state legislatures force them [i.e. our colleges] to admit thousands of students who aren't qualified, who haven't the ability to profit from college," writes Kemeny [1964] (p. 9).

One should come to university in order to learn, and not to fail. And the only way to do that, since more means different, will be by having a variety of courses at different levels. By providing a different set of courses, we could cater for the larger variety of students. But governments and states, although they have increased student numbers, have not increased the funding to universities in proportion, so although we can absorb the larger number of students, we mainly do so by enlarging existing classes. This is quite practical to do, but the net result is that we can devote less time per student, rather than more time, which is obviously required with weaker students. In other words, more teachers per student are required rather than less; in reality, we are getting fewer teachers per student. For instance, the student/staff ratio in England changed from 8.2 in 1972/3 to 11.6 in 1990/91.

Incidentally the student/staff ratio being 11.6 does not mean that classes consist of at most 12 students. This is because staff include administrative and library staff, and university teachers spend a considerable time involved in research, development and administration. I regularly face classes of 40 to 60 students, for instance, although at my university the student/staff ratio is 19 to 1.

With so many students coming into the university, many of whom have no interest, desire or capacity for the study of mathematics, we are naturally encouraged to put most of our energies into simplifying the courses and helping the weaker students. After all, they are by far the largest in number. Surely it is not economic to put any special effort into helping the very clever students? I shall argue strongly against this view.

2.5 Attainment levels

In order to discuss standards we need in some way to classify intellectual attainment. I suggest the following approach, in which attainment is measured by considering the following levels: Abstraction level, Definition level, Techniques level, Proof level, and Ingenuity level.

(Other taxonomies are described in Griffiths [1974], Chapter 20.)

Abstraction Level

Level 1 Handling numbers, starting with whole numbers, fractions and decimals.
Level 2 Concrete objects like polynomials and 3-dimensional vectors.
Level 3 Handling symbols, manipulating equations.
Level 4 Handling theorems, like the mean value theorem, where one needs to check certain condition are satisfied before drawing conclusions.
Level 5 Handling abstract concepts such as groups, rings, fields, metric spaces.
Level 6 Handling concepts like topologies and categories, which are even more abstract.

Definition Level

Level 1 Definition by example.
Level 2 Definition by geometrical or physical intuition, for instance, area under a curve.
Level 3 Precise definitions, such as definition of the cross product of two vectors in 3-dimensional space.
Level 4 Harder definitions, like recursive definition of a determinant, product of matrices.
Level 5 Precise definitions, but of harder conceptual difficulty, such as definition of injective map, linear independence, or limit, or supremum.
Level 6 Even harder definitions, such as uniform continuity, Lebesgue integral.

Techniques Level

Level 1 Ability to calculate with numbers.
Level 2 Ability to substitute values into equations.
Level 3 Manipulation of equations.
Level 4 Algorithms such as Euclid's algorithm.

Level 5 Techniques for differentiation and integration.
Level 6 Harder techniques, such as solving differential equations, Laplace transformations, or using recurrence relations to approximate.

Proof Level

Level 1 Proof by authority.
Level 2 Proof by example.
Level 3 Proof by physical or mathematical intuition, or analogy.
Level 4 Proof by correct manipulation of symbols.
Level 5 Proof by reasoning, with some gaps and intuitive steps, such as is common in Euclidean Geometry. Proof by induction or contradiction.
Level 6 Proof by use of definitions and axioms, and constructing a web of interconnected results.

Ingenuity Level

Level 1 The application of a standard method.
Level 2 A small adjustment of a standard method.
Level 3 Ability to translate a word problem for instance, or a problem of mechanics, into mathematical form.
Level 4 Ability to adapt and or combine one or two known problems to solve a problem in a non-obvious way.
Level 5 Even more ingenious adaptation and application of results in a non-obvious way. A solution which requires a good idea.
Level 6 Able to think of and prove one's own results.

The gaps between adjacent levels are not of equal difficulty; generally the higher the level, the harder it is to proceed to the next.

It is a mistake to learn only the higher level definitions, methods of proof etc. It is always valuable to be able to understand mathematics at all the levels.

I shall use these levels of attainment to explain my views on the attainments of incoming students more precisely, and also subsequently in the book, on several occasions.

2.6 Standards

The whole topic of standards is complex, because students of today may have studied other subjects and may be better equipped in many other ways than their counterparts some twenty years ago; certainly, for instance, students are much better in handling hand calculators, computers and statistics. They are more assured and can handle social situations more easily, they have travelled abroad and know a great deal of the outside world.

But when I say the standards have dropped I mean explicitly a loss of facility in handling those techniques that are useful for understanding mathematical subjects like linear algebra and calculus. For instance, the skills of manipulating algebraic expressions and solving equations, the ability to handle the trigonometric and logarithmic functions, and also the concept of proof and generality, and the ability to recognise curves such as ellipses or circles or parabolas from their equations in cartesian co-ordinates. Determination is an important quality that has deteriorated over the years, that is, the students are less willing to struggle to find a solution, preferring to give up after only a brief attempt at solving a problem, and seeking the solution in a crib. Students seem to perceive a non-standard problem more as a chore than as a challenge.

It is the concept of proof which seems to have disappeared from the school syllabus. When I was a school student I studied Euclidean Geometry, from the age of 12. It was a matter of proving theorems from axioms and definitions. I and my contemporaries had a very good idea of what a proof entailed. But that is not the case with the present day students. More and more it seems as if proofs given at school are given as a matter of form, as if they do not really matter, and nobody is really expected to understand them and certainly not to know them. So when students come to the university and find out that proof and definition are very important parts of the course, they are at a loss.

There are also fewer problems, and most of the problems that appear in the new text-books are standard, in the sense that students need only to look at a worked example in the book and make only very slight adjustments to solve the problem. This was commented on by the LMS [1995] report discussing the results of a test undertaken for incoming undergraduates as follows:

"The lack of success on the third part of this question underlines the dangers of a system which

(1) sets questions which lead the candidate, step by step and
(2) rewards superficial knowledge of those items listed in the 'levels' of the National Curriculum.

Again, there is a substantial drop in success when the question requires additional, unsignalled steps."

If I compare the students coming into the university from schools of today with those of some 20 years ago, I notice that proof level is much lower, and so is ingenuity level, because in the past questions requiring ingenious solutions were common. In terms of the attainment scales described in §2.5 above, I have made a subjective judgement of the attainments of the students coming into the universities (Fig. 2.1).

Level	Abstraction	Definition	Techniques	Proof	Ingenuity
20 years ago	3	2	3	4	3 or 4
Today	2	2	2	3	1 or 2

Fig.2.1 Subjective judgements of student attainments

Thus, according to me, school students entering the universities some 20 years ago were exposed more to non-standard problems, and had a better concept of proof than those of today, and had a higher understanding of abstraction.

2.7 What about the smart students?

Does the fact that the students entering university have a lower standard of achievement than before mean that we must have simpler courses at university than we have had before?

Unlike many European countries (which have till relatively recently restricted university education to an elite), the United States and Canada have for some time had more elementary courses at the universities. These are described, certainly as a joke, and yet with some truth as well, in Stephen Leacock's description of the new education. Thus he speaks to some student who tells him that he studied Religion, Turkish and Music, because "they fitted in well." It turned out that he meant that they were time-tabled one after the other, and that left the afternoon free for his own pursuits.

An essential feature of the modern university, as I understand it, has always been excellence. At the university the aim has always been to extend and retain knowledge and to train students to reach the highest possible standard of academic excellence. If a university can not do that, then it cannot claim legitimately to be a university. It is true that some world renowned universities have produced plenty of mediocre graduates, but that is not the aim, the aim is to produce outstanding graduates.

One reason why it is particularly important to produce outstanding graduates at the university is because after that there is no further recognised training. It is the end of the line. From then on you are expected to go out into society and help produce new ideas and new technology, to contribute, rather than to learn. It is of course possible to add post-university universities, but there must be a limit to how many years a person can continue to study without producing something of value to the community.

We can drop our course standards drastically, and in this way accommodate the problem of weak and ill-educated students. But if we do so we fail in one of our most important commitments, to produce some outstanding graduates, capable, imaginative, with a wide body of knowledge and skills.

Some people say we shouldn't worry about the smart students. They can always manage. However, it is not true. The smart students pass all right and cope with many difficulties brilliantly, they will easily outperform their peers, but if they are truly to become great scholars, researchers and people of high ability, they need plenty of help along the way. "Look after the weak, and the strong will look after themselves," is not a sensible motto.

We must look after the strong students. At the same time there is a large number of students of lesser ability and lesser knowledge who are coming into the universities, and we must develop suitable courses for them too.

2.8 Conclusions

My conclusions, are, as already stated, three in number:

We must not neglect the brighter students, and must give them demanding courses.

At university we must provide a number of courses to suit a variety of student abilities and interests.

Schools must provide the sort of course that will extend and develop the most able students. For these students it makes sense to argue that the school should be designed to fit the student for their subsequent university courses. If the universities provide a variety of courses perhaps the school can go back to having the aim that the school course should fit the student for his subsequent university education, providing courses of various levels.

These recommendations are in conflict with the Higginson argument as quoted in §2.3. The Higginson argument is if only a small fraction of students are going to study a certain subject later on in life, then it is not worth while providing courses with deeper material so that the student can use this material to build up competence in the subject. Instead the idea is to take only those parts of the subject that are utilitarian and of relevance to the majority of students and study that instead.

This argument is in my opinion flawed. For using it, there is no point in studying art, music, theatre, poetry, writing, philosophy or indeed anything, because only a small fraction of students will ever go on to seriously study these subjects. Since most subjects require cumulative study, the Higginson argument is an argument for ignorance.

3. Aims

Preview

Government, lecturers, educationalists, and students, all have important parts to play in a university, and of course they have different aims, which must be considered in turn.

I should perhaps explain that by educationalists I mean people whose main interest lies in getting knowledge and ideas across. They put less emphasis on the content of the course. They look at the amateurish approaches of the current lecturers and feel that there are better ways of teaching. They are beginning to have an increasing role in universities, especially as newer technical methods of teaching using interactive computers and the internet become more and more important.

3.1 Education and employment

It is possible to maintain that education is itself a good, and worthwhile pursuing whether it leads to employment or not. Nevertheless, for most, for students, for governments and for educationalists, improving one's employment prospects remains an important aspect of university studies.

Education and employment have always been important. A friend of mine who went to school in the 40's, remarked that when he studied at school he had numerous calculations to do. "Once you think of it," he said, "you realise that I was being trained to be a clerk, capable of working out things like 143 items at 3 shillings two pence and a farthing each." However, by then, the need for clerks had receded, and the schools had not yet adjusted. He was being taught something quite redundant. (With this in mind perhaps it is worth thinking of whether we too are teaching our students material which no longer is needed.)

When discussing employment one must consider the effects of the computer age. An enormous increase in unemployment has been produced by the computer. There are two important and obvious trends affecting our society due to its dominance.

Firstly because the computer can very quickly handle many routine tasks which previously required well-trained staff, properly trained workers are now no longer required for these tasks. This indicates that one needs less education for many of these jobs, and of course there will be only a few of such jobs.

But then, on the other hand, if people are to have jobs, they must be able to do things that computers can not, and that means they must be even better educated, even more sophisticated than ever before, capable of dealing with difficult tasks which are not easily computerised. In brief, we need more education.

This need for more sophisticated and well trained workers has now had an effect on the demand for universities and this explains in part the increase in university enrolment. However nowadays it is not sufficient to be well-trained and skilled to have a job, because we do not know which are the jobs that will be replaced in the future.

In my own life-time I remember very vividly three different jobs which required considerable skill, training and ability and now have lost their importance. The first is that of a lathe operator, certainly a skilled job, that required considerable effort and capability to master, and it was always thought that with these skills and knowledge one would always have a job. Yet with the event of automatic lathes, an enormous number of people became redundant.

The second outstanding job was that of secretary. Secretaries could travel the world and get a job almost instantly. That is not the case now. Such skills as they had are no longer so essential. The word processor has reduced their importance.

The third job was that of a printer. At one stage printers commanded huge salaries, and the job was difficult, intricate, and required great skill. Nowadays the journalists type their own reports (they certainly do not have secretaries); the electronic versions are automatically and speedily type-set, and the printing takes place almost automatically.

All these jobs I mentioned required considerable skill, training and capabilities, were much in demand, were considered the passport to a permanent and lucrative career, and yet are now much reduced in importance and need. Most of this has been caused by new technology, particularly the computer, which has also eliminated many middle-managers.

One wonders what other category of skilled workers will disappear. I suspect that translators will be the next to go. We are a long way from good computer programs that can translate at the moment, but in ten years time things will probably be different. If mathematical programs like Mathematica are enhanced and are used to a much greater extent, and if the use and provision of multi-media and internet facilities increase, it is possible that university lecturers may also become redundant.

The event of the computer means that people can now solve extremely difficult problems with less knowledge than before. There is consequently no longer such a need for well-educated people. People can be trained in an afternoon to do a job which previously took years of training. I remember the days when I would go to my car spares expert, and ask him for a spare. He would glance at the old one, and say such things as "That is no longer available," and then fumbling through some odds and ends in a box, "But, perhaps, yes, I have this one, which you can use instead, if you file off that flange." He had an immense amount of knowledge, built up over many years.

Nowadays my spares expert insists on the model number of the car, the model number of the part, looks at his computer screen, and almost instantly finds the part on his list. The advantage is that he has not had to have any training other than how to type in the correct numbers. (Somewhat exaggerated and simplified to make my point.)

It is a sobering thought that most people throughout their lives will need to do a number of entirely different jobs. The days when one learnt one skill and was able to support oneself for the rest of one's life with those skills may be over. Society may need to have people who are extremely flexible, and are able to re-train and do other jobs if and when necessary.

It is for those reasons that it is important that a university training should not be directed to some particular skill which has an immediate application or need today. A university course must involve a great degree of learning which trains adaptability and flexibility.

3.2 Aims of society in supporting the university

The aims of the university are varied and many, and depend of course on what point of view we are taking, whether it is the government, or the individual student, an employer, an academic, or an educationalist.

If we begin with society and the government, there are several aims that are of importance. Society wants the university as a status symbol. Towns are very anxious to get a university. It makes them seem more important and valuable. It is part of culture, and just as governments like to have a good national theatre or ballet or art gallery, so too they like to have a good university. A university town is a cut above the others.

Then too if the university has something special or unique that is also valuable. Thus a university with a world-famous algebraic topologist is regarded as an asset. It could just as well be a world-famous number theorist, or anything else, just as long in some way the town or government has the best in the world, so as to boast.

Governments insist that the universities be centres of excellence, at any rate, they like to call them so. They are loath to call their university a training centre, even if that is what it is.

Of course a university brings practical advantages as well. With a large number of students and staff requiring housing, and supplies, the university brings more jobs and prosperity to the town. Indeed in many cases the university becomes the town's largest employer. The university is likely to provide a large number of well qualified graduates, who will fill jobs required by local industries. The hope is also that the university will bring expertise which will enable local industries to solve their problems, and bring further jobs and prosperity to the town and region. This co-operation between university and industry is further facilitated by the introduction of science parks.

Governments hope that the university will serve as a centre of learning and scholarship, that learning will be preserved and be available as a resource. For instance, the university is likely to have an expert, who suddenly is of the greatest value to the government or the town. Thus experts in the Middle-east became of great importance during the Gulf War. In the lightning changes of politics, who knows what unusual language may turn out to be of vital importance, and where else than at the university will there be somebody to speak and write the now important language?

In the same way, suddenly a discipline previously thought to be unimportant, will come to the fore, and academics who seemed to be there just for cultural reasons, will be called upon to teach students who will then be filling vital jobs. All of these items are almost certain to pay off. But there are also the apparently impractical and unpromising directions of research which will produce enormous benefits, often many years later.

There are plenty of such examples. The Crooke's rays for instance, which led to the television tube. The development of precise ways of expressing Algebra, which led to, among other things, useful programming languages. Ideas in Logic, which also helped develop computers, and also indicated their inadequacies. Number Theory, perhaps the purest subject in pure mathematics, now provides practical codes.

So for society as a whole, the university means a direct source of employment, a supply of educated workers, a technical tool to solve industry's problems, a source of suddenly vital knowledge, and maybe some brilliant new earth-shattering development.

Recently governments have used universities to reduce the apparent unemployment. By making people study instead of being on the dole, governments achieve two things: they are reducing the apparent jobless total (important politically) and at the same time training a large number of people ready for the new jobs if they should arise. Since in Sweden one takes a loan to study, and each person must repay the loan, the Swedish government is thereby also spending less on benefits.

The hope too is that the country will be ready for an upswing, and will not be held back by lack of qualified people. Also it is felt that in the future there will be jobs only for highly skilled and trained people.

There is a strong feeling that the universities are essential to a developed society's continued prosperity. As Derek Bok, president of Harvard University, put it, "More and more, therefore, the United States will have to live by its wits, prospering or declining according to the capacity of its people to develop new ideas, to work with sophisticated technology, to create new products and imaginative new ways of solving problems. Of all our national assets, a trained intelligence and a capacity for innovation and discovery seem destined to be the most important." Similar ideas are expressed in most countries by almost everybody who is concerned with education.

Also many businesses now require workers who have been trained at university level. In many cases these people need even further in-house education. This is due to the detailed refinement that has gradually been achieved over the years, and the complexity of products. Also, it is not sufficient to be the best of the companies in a particular country, because companies from all countries of the world can compete in every other country. Almost every good product is sooner or later marketed all over the world.

It would seem to be an obvious point that universities should be constantly checking to see what employers are looking for in their new employees. However, on the whole universities do not bother. This is in part due to the fact that they do not have the time nor the facilities to check. A notable exception is the report of McLone [1973] which is excellent, and shows clearly what can be done. McLone's investigations should have been repeated every five years or so, and now the original report is very much out of date. Indeed, industry itself has changed enormously during the last quarter of a century, with manufacturing industry on the decline and service industry increasing.

Nevertheless it is well worth quoting McLone's summing up, since I believe it to be equally true today:

"There remains a demand from industry and commerce for graduates with a high level of mathematical ability. Correspondingly, there is little demand for academic pure and applied mathematicians, but rather for graduates with a sound basis of mathematics who appreciate the wide range of applicability of their subject, who can translate industrial/commercial problems (of many forms, not simply technical) initially expressed in non-mathematical terms into a form amenable to mathematical treatment, and subsequently to re-express the result in a form readily assimilated by non-mathematical colleagues. The ability to apply mathematics (rather than develop new areas of mathematics) is of most importance to the majority of employers and graduates alike."

Although it is of some importance to notice the needs of industry, it is not sensible to design courses too slavishly on these needs. By the time the student graduates, those jobs may no longer be available, or else developments may have led to industry requiring other knowledge or qualities.

3.3 Aims of students, educationalists and academics

If governments, society and towns have special reasons for supporting a university, so do the students. They certainly regard university as a chance to get a qualification which will entitle them to a well-paid job. Some of them may be scholars, whose main aim is to pursue knowledge further. Some will have come along because that is current: one goes through school and then continues on to university. Some may regard their studies as a chore, an obstacle race that must be performed so that they can enter the profession they are interested in. Some will realise that they are being equipped with essential knowledge: nowadays every job is so specialised that one cannot hope to work without considerable academic knowledge.

When I studied at university many years ago, some of the young women students and their parents regarded the university as an excellent marriage broker, with the chance to educate yourself and also meet a young man (with an education and hence good prospects) as well. Thus there have always been other aims extraneous to academic aims, and there will probably always be such aims.

Getting to the right university can make a difference too in later life. For instance, a graduate from Cambridge University sounds very much more impressive than a graduate from Keele University. Part of it is snob value, but some if it may have a sound basis, since on the whole, the more prestigious universities can attract better students right from the beginning, and then can insist on higher standards for the whole course.

In general, going to Oxford or Cambridge is a distinct social and financial advantage in England. Their graduates are disproportionately represented as government ministers, as members of parliament and other important positions. Indeed, going to the right place can mean many useful contacts for the rest of your life, a way of easing your way for ever more.

It is also the case that studying may be a person's best short-time financial option. Some students in Sweden are not really all that interested in study. On the other hand, they are out of a job. They can get a loan which will enable them to live relatively comfortably for at least a year. Unemployment drives them to the university.

So the reasons that students see as valid for a university are certainly very numerous and varied. It is a mistake to dismiss their aims as frivolous. Many students are as well aware as academics that they need a deep and demanding course. It is only those who are forced (quite mistakenly) to study mathematics simply to qualify for the course they are really interested in who tend to resent the efforts they are forced, quite needlessly, to put in.

It is important for academics to realise the many possible motives why students come to study at university. One reason for this is that the way to stimulate students is to take their interests and relate them to the course. The more one knows about one's students the easier it is to say something which seems relevant to them. This is nothing esoteric or difficult. It is something we do every day when we speak to our friends. We speak differently to different friends, realising that what will interest one will bore another.

Do not forget that what every student is interested in the examination. To say to the students that a particular topic is going to be of importance in the examination is not teaching a love of the subject, but it at least will help them to master the course and should not be despised.

What about academics? Many academics still think that the university is and should be regarded as similar to an elite athletics club. Most normal people can not possibly reach the high physical standard demanded by top class athletes. To subject a group of normal people of various ages and strengths to the sort of training required for competitive athletics will lead to exhaustion or even death. To adopt an easier regime will result in athletes who can in no sense compete at a high level.

The same applies to universities. Teach students at a rate and standard that the average can cope with, and you will not end up with the champion graduate who can compete world wide. Formerly we were able to consider the elite approach. Students were highly ambitious and had studied widely and deeply at school. We cannot do that now. As we argued in Chapter 2, the vast expansion of numbers at the university means that we do not have an elite.

There are the academics who regard the university as the repository of knowledge, a place for the conservation of all the brilliant ideas that mankind has had. Not only must this knowledge be preserved, it must also be transmitted to the next generation, so as to keep it alive and flourishing.

For those who are interested in research, that is the most essential role of the university, to push forward knowledge in all directions, whether there is any practical value in so doing or not.

In the United Kingdom the government assesses research, and departments which do not get a high rating receive less financial support. This of course makes research even more important for academics as a whole.

Many of the academics are also hoping to educate a number of very clever students, who will continue to preserve the discipline, and also to advance it. It is these superior students who are in danger of being swamped and lost in the more democratic universities of today.

The academics are not solely concerned with teaching; the studying of mathematics, research, administration, all require their attention. Many of them do not see teaching as their main role. A group that is mainly concerned with teaching are the educationalists.

This is a good point to distinguish between training and education. Training consists mainly of learning what to do. A good example is the rule of casting out 9's. Here for instance if we work out say a product of two whole numbers, we can add the digits of the multiplicand and the multiplicator and then multiply them, checking whether we get the same result modulo 9 when we add the digits in our answer. This provides some check of our calculation being correct. Explaining and giving drill on using the rule is training, explaining why the rule works is education.

Obviously both types of teaching are important. The two also reinforce each other. Thus a medical doctor needs to be trained on emergency procedures, say if somebody's heart has stopped. This training must be very thorough, so that the necessary procedures can be carried out immediately and faultlessly when needed. On the other hand, without a knowledge of the heart and the circulation it would be less easy to master these procedures, or to have any idea of what one was doing.

Universities will probably need in the future to provide more training. There are so many technically difficult and demanding jobs where trained workers are needed. For instance, people trained in certain computer programmes and procedures can get a job immediately, so much so that some students of computing instead of completing their university education will leave early after having obtained that sort of training. But the university must still provide education for other students.

The educationalists are interested mainly in the students' learning, and are keen for the lecturers to organise the students' learning with activities and tests. They go more towards directing the students' studying, more towards training rather than education. The mathematicians are mainly interested in creating good mathematicians, clever, innovative people who appreciate mathematics. They expect well-motivated students who have already learnt a great deal and are determined to learn as much as possible. They aim to explain the ideas clearly and well, but they expect the students to do the learning themselves.

Whatever the differences, most agree that whether the student is clever or not, there is the aim that the student must become independent. Thus students must be able to teach themselves. It is one of the theses of this book that unlike in the past when students learnt to teach themselves merely through coping with difficult tasks, they need to be taught the art of learning in a more direct and deliberate way.

For a university to flourish and succeed, it is necessary for all these disparate ideas and aims be achieved, so that the university can go on receiving the support it does from students, academics, and the community at large.

It is important however to notice that students should on the whole be successful in learning when they come to university. They come to learn, and we should organise things so that they do learn. Excessive failure rates are not satisfactory.

There is yet another important body of people who determine pass rates. These are the professions, who like to maintain high standards as a means of keeping the number of people in their profession low enough to ensure high wages for themselves and their friends who are already in the profession.

So at some universities students who have no desire, need or talent are forced to study a mathematics course, which acts as a filter, only allowing a certain number of students to go on to study medicine or business studies or what have you. In this case it is seen to be an advantage to have a fairly high failure rate. This function I must dismiss as unworthy of a university: fundamental to a university is the concept that students should come to learn, not come to fail.

3.4 Why learn mathematics?

It is becoming more and more essential to study at university as the competition for jobs increases and the dearth of jobs may very well mean that a person in practice has to choose between continuing to study or to continue to be unemployed.

But why should one study mathematics?

If one is interested in courses such as engineering, physics, chemistry one has no choice. These subjects require as part of the discipline a considerable knowledge of mathematics. Subjects like economics, biology, psychology and sociology are increasingly requiring mathematical skills as well.

It is perhaps hard to say why mathematics has proved to be so successful in being applied to the world (see Wigner [1960] for an interesting discussion), but the fact is that it is certainly very effective.

For instance, nowadays much of a new aeroplane is mathematically designed in advance. There is also experimental work of course, but it would be quite uneconomic to keep on building prototypes. Planes tend to fly successfully, and the host of new and faster, quieter and more economical planes bear witness to the effectiveness of the mathematics. For instance, James Lighthill was involved in the development of supersonic aeroplanes and in making them quieter. His mathematical arguments also made the River Thames cleaner, and surgical operations safer.

There is also an argument for studying mathematics for the purpose of developing one's mind. The idea of developing the mind and learning has two main aims. There is firstly intrinsic merit in learning for its own sake, in developing one's ideas and abilities. Then in practical terms, armed with these general abilities one will, in any instance, be able to apply the same skills to good effect. The type of deep and precise thinking needed when studying mathematics can often be usefully employed elsewhere. It seems for instance that a person with a mathematics background can more easily learn the economics required and then apply the mathematics, than it is for an economist to learn the needed mathematics and apply that.

Mathematics is part of the culture of our times, and all of us need some understanding of it. It is difficult to sensibly discuss different plans of action involving complex issues like traffic plans or hospital funding requirements without having a reasonable understanding of mathematics and statistics. But for many, mathematics has its own intrinsic interest, and it is worthwhile studying for its interesting ideas and results.

However, studying mathematics is not suitable for everybody. This fact is as simple as the statement that not everybody can become a successful marathon runner. Not everybody can learn mathematics. Not everybody can be academic. The evidence is in the large number of people who struggle very hard and get nowhere. Yet this simple statement is not accepted by many politicians.

One politician put this vividly. "Look," he said, "people like you (talking of those who were protesting that the educational system could not take so much expansion) used to say that it would neither possible nor desirable to teach everybody to read. Yet we did it. People used to say that it will not be possible to teach everybody to gymnasium standard[5]. Yet we do it now. In the future it will turn out to be usual for the average person to get at least two university degrees, since it is unlikely that the one degree will suffice throughout life. And that will also happen."

As the politician says, in Sweden most people now go on to the gymnasium. It is furthermore required that every student at the gymnasium reaches the same standard in mathematics, Swedish and English. The schools have reacted by improving the teaching, by getting better books, and by lowering the standards. Nevertheless about a quarter of students fail to reach these lower standards.

I once heard a radio discussion in which a school teacher tried in vain to explain that this policy was having a dreadful effect on the self-confidence and the well-being of the students who could not make the grade. The education minister he was talking to refused to accept this. In the view of the education minister, everybody must have these basic abilities in mathematics, Swedish and English otherwise they can not go on to university and would be unfairly deprived of the opportunity of further progress in life.

But people do differ. There is a tendency to deny that these differences exist. I remember saying in Sweden once that I had a large number of "stupid students in my class." A hush fell over the group of people I was speaking to. They were all visibly shaken. One of the hardier members of the group, who managed to recover more quickly than the others, explained that in Sweden one did not use such words. One could, if one had to, talk about "less talented students."

Now I know of course, that ability is not one dimensional, and it is quite possible for people to be very weak at mathematics, while at the same time they may be brilliant at other things. And I know that a student may be weak now and may improve dramatically later. So the use of the word "stupid" can not be entirely defended. But the point of the story is that we have all been conditioned into denying that people have different abilities.

[5] The last three years of schooling before university.

3.5 Conclusions

Students do differ, and they differ markedly, and there is no use denying the facts. My conclusion is that we must handle both the clever students, and the weak, and of course those in between. We need to enable all our students to make the very best of the abilities they have.

However, the universities have not been given funds so as to carry out new courses. Presumably we are expected to handle the students all together, in other words, aerobics for all. However this is for all the reasons above unsatisfactory. We need a greater variety of courses and we need to fund them.

School reforms have improved the treatment of the less able students, while the more able students have been neglected. The schools have not yet found a satisfactory way of ensuring bright students attain their potential. This should be an important goal for schools. The universities must ensure that they develop, to the highest possible level, the capabilities and knowledge of the best students.

4. Universities and Government

Preview

In a number of countries governments and legislators are taking more control of universities. Since they regard the university as an important asset and since they often foot the bill, this is natural. It is important to note the effect they have on entry standards and on pass rates, on university staff and their conditions.

These effects are so extensive and far-reaching that here we have yet another reason why academics must take a greater interest and be more active in discussing these matters.

4.1 Government requirements

Again each country has to be considered on its own. In this chapter I will take up the situation in Sweden, as an illustration of what I mean. But it is likely that the influence of government everywhere is equally significant on matters such as number of students, entry requirements, the effect on teaching methods, the demand that universities justify their work, and the percentage of students who are expected to pass.

Sweden is of special interest, because it is a country in which education is regarded as important, the government exerts strong control and the precise details are clear-cut. Many of the education decisions are naturally enough politically rather than educationally motivated.

The Swedish government decides the entry requirements. The more students the university accepts, the more money it gets, up to a certain ceiling determined by the government for each university. This leads one to try to accept as many students as one has space for.

The basic requirements are either a pass in the three year gymnasium program or else one should be 25 or over, have worked for at least four

4years, and have certificated language skills in English and Swedish. (This last condition is soon to be dropped.)

In Sweden the government is determined that the failure rate will be low. For the year 1997 the Swedish government paid about SEK36,000 for teaching for a full student year, and SEK33,000 for each student who passed all the courses in that year. In other words, the university could virtually double its income by making sure the students pass. The idea behind this funding is obvious: it is that standards will remain unchanged, but universities will now have a large incentive to improve their teaching.

Improving the pass rate and keeping to the same type of examinations and examination conditions (e.g. closed book examination, same content, equally difficult questions) is difficult. Improving the pass rate while relaxing some of these conditions can be done without any difficulty. One can ensure a better pass rate by a whole range of measures.

Indeed, in order not to be too obvious about what one is doing, it is best to use as many measures as possible. For instance, one can begin by reducing the course content by 10%, which is hard to notice, especially in Sweden where there is no external examiner to maintain standards. Reducing the course by 10% however makes a tremendous difference. Not only is there less material to learn, there is also more time to learn the reduced amount of material. Presumably the 10% of material left out will be the material that used to be covered last, and consequently the most difficult material in the course.

Then one can reduce the degree of abstraction and generalisation. Instead of having proofs one can teach by example alone.

Then one can make the examination easier, by avoiding any non-standard questions, or reducing the number of parts required per questions. For instance, if previously one would set a question which entailed four parts, one could set a question with only three parts.

It is quite common to give a specimen examination and a set of answers. One can make this specimen examination very similar to the actual examination. Or one can tell the students one or two of the questions that will appear on the examination.

At the same time, one can drop the pass mark. Or perhaps, decide on what seems a good pass-rate, say 80%, and pass the top 80% as judged by their mark in the examination.

One can also split the examination up into several smaller examinations. Or one can arrange to increase the number of opportunities to pass the exam.

So there are all sorts of methods which can be used to improve the pass rate without maintaining standards. I am convinced, that although academics are honourable, the standards will be lowered so that the pass rate will improve drastically. Of course there may be some universities which resist, but since they will then receive less money, they will soon disappear, or else learn the error of their ways.

Perhaps I should make it plain. I am not advocating that universities lower their standards in this fashion. Indeed, I am against a lowering of standards. All I am saying is it is inevitable with government direction on who can be admitted to university, and with funding dependent on passes, standards will automatically become lower.

The Swedish government laid down the following requirements for universities for the budget year 1997.

(1) Firstly, to produce work at a high international level, for undergraduate and postgraduate education, as well as research.

(2) To interact with the community at large, both by contributing to development and research of commercial businesses, and also by informing the public at large what they are doing.

(3) They were required to intensify their work on teaching methods, developing new methods, and were supposed to offer their staff courses on teaching methods.

These requirements were unreasonable: the university staff could not possible achieve so much simultaneously. The following section will explain this in greater detail.

4.2 The lecturers

The university used to provide an attractive job, reasonably well-paid, interesting, with plenty free time, and with sufficient flexibility to enable one to pursue a life of learning and research. There was always intense peer pressure. The unsuccessful researchers of course benefited less; they had less chance to travel, they were given more administration or teaching duties, they were not respected. If the academic was not too sensitive, and did not mind being regarded a failure in the eyes of colleagues, he or she could still have a pleasant life, and could expect, once employed, to have a job till retirement.

However, this happy view of the life of an academic has changed, in view of the deterioration of conditions, the uncertainty of permanent employment, a salary which is less competitive, extra demands with more students to teach, and also students of lower intellectual capabilities. The successful Ph.D. graduates discover that having spent at least 6 years getting a degree, that they are expected to spend years hunting for one temporary job after another, living at much the same standard as a student, in the hope that they can publish enough papers and accumulate enough teaching experience to obtain a permanent job. They are unlikely to be able to choose where they want to live, but will need to go to wherever in the world there is a job. The profession of an academic is no longer idyllic.

Governments are also introducing demanding rating systems, rating staff both as to research and teaching, and giving less money to those departments who rate badly. Consequently the job is no longer as permanent as it once was. Lecturing is becoming less attractive.

Indeed, a university lecturer has two full-time and deeply competitive jobs simultaneously: the first is to do research and stay abreast of one's field, the other is to teach students. Few realise just how hard it is to do either, and how much work is involved.

The chances of promotion for lecturers are small—indeed it is difficult even to get a permanent job. However, when the present large group of lecturers (recruited mainly in the 60's) retire, it will not be long before it is hard to staff universities.

As governments insist on more and more research and more and more teaching from university staff, and as conditions inevitably deteriorate, those willing to go into an academic career will become fewer and fewer, and the quality of lecturers will inevitably decrease. So far this has not occurred, because of the huge influx of brilliant mathematicians from China, the USSR, and other eastern countries.

Thus the richer countries can and have benefited enormously at the expense of poorer countries. This has also happened in the case of war or persecution.[6] But relying for one's mathematicians on other countries is not a sensible procedure.

---

[6] Both the United States and the UK benefited from an influx of brilliant German academics prior to and during World War II.

Problems can also arise with the older foreign mathematicians. They often find difficulty in mastering a new language, and thus do not have the skills required to present mathematics as well as they should to the weaker students. (The better students probably suffer less, but they too would benefit from people who can express their ideas well.)

I myself, as a foreign lecturer in Sweden, know that my lectures lack precision, fluency, and variation, style and interest due to my somewhat restricted Swedish. Because mathematics has such a limited vocabulary, one can do a passable job, but mathematics as it is suffers from a lack of good expositors, and the position will surely deteriorate if students do not have at least some good role models.

The much larger classes are of course having another effect. Mathematicians desperate to continue their research work find that they lose a disproportionate amount of time helping individual students. Consequently their best strategy is to give as little help as possible. Since the present students are in fact weaker on the whole, they really need more help. Government pressure and demands are tearing lecturers in two directions.

Thus university lecturers do not have the time and the resources to improve their lecturing. The calculation below gives an indication of how pressed Swedish academics are.

In Sweden in 1997 there was a ceiling of 400 contact hours per year for lecturers (other categories such as professors and adjuncts have other regulations, but I will stick to lecturers for the purpose of my argument).

Based then, on a "normal" work-load of a lecturer with 400 contact hours, we have the following rough estimate (Fig. 4.1), done by one head of department.

The teaching was calculated at 400 contact hours, and this total was multiplied by 3 to allow for preparation time and any individual help given to students.

Now not everybody teaches up to the 400 contact hours level (in 1997/8 I had 353 contact hours for instance), but still the time left for research, co-operation with industry, improving lecturing and informing the public of the new results which may be relevant to them (that is, carrying out the essential duties of their employment according to the Swedish government as described in §4.1) is insufficient.

General administrative duties	100 hours
Self-development	173 hours
Teaching, including preparation and completion work afterwards	1200 hours
Examination marking, committees, supervision, development of courses	300 hours
Total[7]	1773 hours

Fig 4.1 Estimated work load of lecturers

It is recognised that the 400 hours of teaching is excessive. Thus in the International Review of Swedish Research in Mathematical Sciences we read that "The Committee thus recommends a drastic change of the academic structure in the mathematics departments. The number of positions as lecturer should be decreased and every lecturer should be guaranteed 50% of his/her time for research" [NRC 1995]. The Committee seemed mainly to think that the 400 hours was excessive because one could not do research. They did not comment that even if one were to restrict oneself to teaching, that being innovative, finding new methods, giving students individual attention, that even then, 400 hours is too much.

There has never been anything in the underlying structure of university employment to encourage lecturers to try to improve their lecturing, indeed, the way to further progress in one's career has always been a good research record. In the words of the old rhyme: "A theorem a day means promotion and pay, a theorem a year and you're out on your ear!" But things are getting even tougher now, especially in countries like England, where the future of whole departments depends on their research ratings.

[7] The 1773 hours total can be more easily comprehended by expressing it in working weeks of 39 hours. It then works out to 45½ weeks per year.

4.3 Summary

It is clear that political decisions have an enormous influence on higher education, on the number of students who enter, on the entrance standards, on the pass rate, on the pressures on lecturers.

Part II General Theory of Teaching

Chapter 5 Teaching

Chapter 6 Study Skills

Chapter 7 Rules of Teaching

5. Teaching

Preview

The previous four chapters have been concerned with the way in which education is organised in general, of the changes that have occurred, of the aims and of the influence of government on education. Now we turn to discuss teaching.

We begin by considering whether teaching can be improved. Then we consider the content of courses. Again I emphasise the need for a greater variety of courses, and also the need to consider what industry feels is of value. This leads me to discuss the dated, but still useful, account by McLone.

It is held by some that we are on the threshold of a revolution in teaching practice, where new technology will have a profound effect. However, as yet these new methods of teaching are not very common, so we still need the standard teaching and lecturing skills now and in the reasonably foreseeable future. It is these that are mainly discussed in this chapter and in the rest of this book.

The auxiliary use of calculators and computers (for instance for students to multiply and divide), has already had its effects on the learning of mathematics, and we discuss these briefly.

5.1 Improving teaching

Teaching can be improved. But one should not be over-optimistic. One can not expect miracles. As yet nobody has come up with an ingenious new method of teaching which does marvels. Stephen Leacock contemplating the future, describes a method of surgically operating Spherical Trigonometry into the small intestine (which was no longer needed in the new Utopia because one derived all one's nutriment from a single daily pill), but of course he was writing as a humorist, rather than as an educationalist. Mathematics is a difficult subject, learning still has to be done by the individual, and possibilities for teaching improvements are limited

As one's first and major aim one would like to give a clear, careful and accurate description of the material to be studied. Such a description is of great importance and value, and is extremely difficult to achieve. Much of the work of preparing a course is in fact carefully working through the material one knows and re-organising it in a satisfactory way. One needs to make the whole fully comprehensible. That it works is shown by the fact that most students find it easier to study with the help of lectures rather than straight from text-books. One friend of mine was particularly skilled in this sort of exposition. There did not seem to be anything original or striking in his presentation, yet everything was included at just the right place. There was the right amount of repetition, and extra help with the ideas came when needed. The rest of us tried but could not emulate his skill.

University courses have traditionally included a large amount of material. In practice it is difficult to cover this material in the time allowed during the lectures. The student in fact is expected to do much of the work alone. To use further class time to elaborate, to give more explanations, to try to simplify further, is quite beyond the available hours, not only for lecturers, who have to invent, prepare and deliver the material, but also for students, who require time to listen to and absorb these ideas. With large classes, for example I have about 60 students[8] in my class, it is also very difficult to give individuals help.

Of course there is always the possibility of reducing the quantity of material covered in the course. The trouble is that each year there is more knowledge, not less, and one is reluctant to teach less than we used to twenty years ago. Many of us feel that this would reduce the value of the course, and is thus undesirable. However, it is a possibility. It would lead to a generation of graduates with substantially less knowledge and skill than the present generation of engineers, and scientists. Could they cope in today's increasingly more technically demanding world? Somehow I doubt it.

Contrary to the comments and remarks I have made about the standards of the students coming from school, there are people who feel that there is nothing wrong with the previous education of these students. (See my comments in §§2.3 & 2.6.) The fault, it is claimed, is mainly due to the

---

[8] This remark received the comment "Lucky you. I usually teach between 100 and 200 students."

university lecturers. If they taught better the students could learn almost anything.

Governments are also convinced that the fault lies mainly with the teaching skills, or lack of them, at the university. Thus the United Kingdom has introduced an authority to assess teaching quality at universities, and, as mentioned in §4.1, the Swedish government made it a point for the budget year 1997 that universities had to intensify their work on development of teaching methods and actively find new methods in teaching both for undergraduates and postgraduates. Lecturers had to be offered courses on pedagogy, and the university was required to account at the end of the year for the developments achieved and the number of their staff that had attended courses in pedagogy. There was thus clearly the suggestion that improved teaching is the main way of making progress.

The problem is reminiscent of the alchemists' search for converting lead into gold; nobody knew whether it was possible or economic, and much time and energy were frittered away.

Some of the new methods seem to be effective, but at the cost of a tremendous amount of work by both students and teachers. One of my colleagues explained with great pride that the new method of teaching he had introduced had meant that the students had been forced to work much harder than ever before. As a consequence they had learnt a lot more. He therefore felt his new method of teaching was a success. I cannot agree. I find it quite natural that if somebody works harder in an effective manner he will learn much more. The whole point of providing instruction is to reduce the work required for learning. If my colleague had claimed the students had learnt more with less work I would have been impressed. But not with more work. That is why it is important to measure or else estimate the work done both by students and lecturers when trying to assess the worth of different methods of teaching.

Personally I feel that there are no radical new methods of teaching as yet which will result in a worthwhile improvement with a reasonable investment in time and effort.

Some of the new ideas under-estimate the value of having a teacher-centred part of the course. Courses taught by a variety of methods, with a variety of lecturers showing their own ways of thinking, are what one should aim for. Teacher-centred teaching has always been effective and is one of the most natural ways of learning.

Much of school teaching has been dominated by the ideas of educationalists. Even in the 1960's Kemeny [1964] complains "...that the 'educationalists' have taken over our schools." "The educationalists seem hardly to care what, if anything, a teacher knows about his subject, they only care that he teaches 'properly', using correct psychological techniques."

Now these same ideas are coming into the university. It is as futile to ask a teaching expert, ignorant of mathematics, to help teach mathematics, as it is to ask an expert in sexology who has never experienced sex to resolve our sexual problems. I am not saying that we can not learn a great deal from these experts in teaching. However, their suggestions lie on the periphery, they do not handle the basic problems in the teaching of mathematics. Even if we carry out all their recommendations, we will not have got much further.

We should certainly put effort into improving teaching, taking and improving the tried and trusted methods. I am talking about methods that do not take much time, and make a difference, which should also be cost effective, for one should remember always that there is little time. Teaching is not the only job lecturers have. In order to prosper or even to retain their jobs, lecturers needs to review, maintain and enlarge their skills, and carry out research.

Attention to small and obvious improvements like the ones I will advocate, will make a worthwhile difference. Many improvements however require a combined effort, the efforts of government, administrators, schools and university departments as a whole, and lecturers: they can not be left entirely to lecturers alone.

Most teachers are satisfactory, if not inspiring. There are however some teachers who need to improve, who get the obvious things wrong. This is probably the quickest and most fruitful point to get improvement. The university should put some effort into finding out who these hopeless teachers are, and get them to improve. It need not be a long or difficult exercise. It will be embarrassing, and some people are too pig-headed and too hopeless to learn. Somehow they must be dealt with. We cannot shrink from such tasks.

It could be argued that far from the university lecturers doing a bad job, they have actually done a very good job, when one considers the lower quality, the lack of determination and preparation of the students, and the fact that minimum entrance standards are decided by others. After all, the study of mathematics is far from easy. It requires deep concentration and determination. Similarly, if teachers of the violin were required to accept anybody who had passed elementary lessons, the pass rate would be very low or else the graduates would be indifferent violinists.

Of course that is a crucial consideration. If a student despite the best endeavours can only obtain an "imitation knowledge" of the subject (I use this phrase in the sense used by Sawyer in the quotation in §2.3), is there any reason for the student to continue? It is true that such a student is likely at least to learn how to carry out procedures, and algorithms with great care, but probably not much else.

Nevertheless there are those students who need their mathematics to help them understand some other topic, for instance engineering. In such a case, the need is more for training in procedures and algorithms rather than in the proper understanding of mathematics itself. In such a case it must be understood that such a graduate can not be trusted to make calculations (other than the standard) using mathematics without supervision. In the past, such students would not have been studying at the university, but now they are.

I conclude that the university of the future is going to have a number of streams simultaneously: a tough, demanding stream where we teach students to be independent and to think for themselves, and other streams where we proceed a lot slower.

There are all sorts of difficulties which arise when one has several courses on the same subject but of different difficulty. There is the problem of prestige courses. There are those students who despite severe limitations in understanding nevertheless insist on studying the more prestigious course. The net effect of this is of course to tend to reduce the standard of such a course, until eventually the weaker students bow to the inevitable, and give up.

Nevertheless, these more demanding courses for smart students are an important part of our commitment to keeping standards and ability high.

5.2 Content

What should be taught? What should the content of mathematics courses be?

This clearly depends on the type of student.

I begin with general students, by which I mean students who do not need mathematics directly for their career in the future, but need some understanding of mathematics in the same way as we need some understanding of politics, economics and first-aid for our day to day lives.

At the moment, at universities in Europe, it is assumed on the whole that students should learn traditional mathematics, such as Calculus and Linear Algebra, Differential Equations and so on.

In North America there are a number of courses which are not the nuts and bolts of mathematics, but courses about mathematics, its significance, history, philosophy and development. People with little mathematical technique have been studying these courses for some time. For want of a better term, I shall call such mathematics "soft mathematics." When I use this term, I do not use it in a denigrating fashion, feeling the subjects studied are of considerable value. It is convenient to refer to the other mathematics as "hard mathematics."

An example of such soft mathematics, is the book by Paulos [1991]. It is clear that the students learn a great deal about mathematics and that the topics taken up are certainly interesting, although the book does not include any really demanding, rigorous mathematics. What the students have learnt however, they understand. The course is of considerable value to these students, and much better than the "imitation learning" of mathematics described in §2.3 in the quotation by Sawyer.

Some "soft courses" are mentioned in a list of alternative courses originally collected and organised for the LMS by Tim Porter, a brief account of which appears in Appendix E.

For students of "soft mathematics" I recommend the following:

(1) A list of typical problems which can be solved with mathematics. The student should knew of the major branches of mathematics are and what they are about.
(2) The student needs to learn to be number-wise.
(3) Using the spread sheet to decide between options.

(4) Some idea of statistics.
(5) Some understanding of proof and logic.
(6) Some number theory.
(7) Some history of mathematics.
(8) A little programming.
(9) Some Calculus.
(10) Some exposure to Fermi problems.

By Fermi problems I mean problems where one is required to estimate something without having sufficient data or information to do it. One needs to use a considerable amount of ingenuity to do these calculations. The name comes after a practice of Fermi of giving such problems when teaching Physics students. (For example, estimate the number of piano tuners in Chicago.) A striking example of this technique is given by Wier, and quoted by Stewart [1998], in which an estimate is given of the number of slaves required to build a pyramid.

Obviously the above list is too large for any normal course, and one would need to select carefully.

The obvious books which can be used to prepare such a course are Courant [1996], Davis [1983], Dunham [1994], Gullberg [1997], Huff [1993], Katz [1993], Kordemsky [1972], Paulos [1991, 1995], Polya [1977], Stewart [1987].

It seems to me that we would very much benefit having a large number of these alternative "soft" courses to offer students who do not profit from the more traditional courses.

Perhaps a word about Calculus. It seems to me that Calculus is such a central part of mathematics and physics that some Calculus should be studied by most students. A very short "soft" course with methods of finding maxima and minima and finding areas and volumes, in which calculation is much reduced through the use of programs such as Maple, is not difficult to design.

It would indeed be a pity if the students of "hard mathematics" do not also study the "soft mathematics", but in practice they do not. Such students do not in general try to understand the overall structure, background, and history of their studies. Consequently they have only a fragmentary knowledge of the subject.

The next important group of students to consider are Engineering and Science students. The topics studied are mainly determined by the students, the engineering disciplines, and the science faculties. It usually includes some single and multi-variable calculus, some vector calculus, some linear algebra, some differential equations, some transform theory, some complex variable, and of course, some numerical methods, some statistics, some operational research and some computation. The student would also benefit by learning some combinatorics, and a knowledge of field theory and error correcting codes would be a good idea. A useful book to consult is "A Core Curriculum in Mathematics for the European Engineer", Barry [1995].

There are some, mainly Pure Mathematicians, who believe strongly that for Science and Engineering students, results should be proved with a fair degree of rigour. The argument is that since students will never again see that degree of rigour in future life, it is best that they see it as students.

Personally I don't think that is a good plan, certainly not for engineers, who on the whole see mathematics as a tool, something to use as and when they need it. They really are not too concerned about rigour. If they have a problem, then they will consult a mathematician, rather than work it out themselves. They are quite content to be told how things work. Techniques, methods, tricks, are what they are looking for. Similarly, physicists are more interested in suggestive ideas, rather than detailed and accurate proof.

For these groups of students, although I do not advocate a theorem/axiom/definition approach, I still feel that it is important to derive and explain most of the results, rather than simply state the results or algorithms without further justification. It is also important for them to be able to use theorems in a rigorous fashion, that is, to check carefully that the conditions of the theorems are satisfied in a particular application.

It also seems to me that it would be appropriate for such students to also study and know the topics described above as suitable for the general students, though lack of time would probably mean this could not be achieved.

However, there are a number of suggestions that the mathematics normally taught to engineers is not appropriate for their needs. Chris Bissell (p365 Burton [1995]) is quite scathing about ordinary mathematics. "...Practising engineers, on the other hand, rarely—if ever—need to carry out such tasks. ... the context in which such 'mathematical activities' are carried out is often so far removed from conventional engineering mathematics that it makes more sense to think of engineers and technologists speaking a language essentially different from that of mathematicians."

However, not everybody thinks so. "It is important if you are in charge of a young engineer and his development that you get him to use his mathematics. If he stops using his mathematics his value to your company rapidly dwindles." (Nils Mårtenson, director of a mobile telephone company.)

For mathematics specialists there are of course other criteria. There is usually much argument among the staff about the courses which should be taught. In general it would be sensible if a wide variety of mathematical topics could be covered. Some calculus, analysis, combinatorics, number theory, linear algebra, algebra, geometry, topology, differential equations, complex variable, computing, numerical analysis and statistics, for instance. Some physics, e.g. mechanics or electricity seems a good idea as well. Often the subjects taught depend on the interests and specialities of the staff.

But some general remarks. There are a large number of standard algorithms and techniques in mathematics. It is essential that these be taught. It has been suggested by some that algorithms are the lowest of the low and that one should really aim higher than a "mere" algorithm.

In fact, the algorithm is very similar to set procedures, a very important part of most ordinary lives. Almost all organisations are run on these procedures. If that happens, do this. It is laid down. People do not need to think out each case for themselves. The collective wisdom of those who have wrestled with these problems has been captured in a collection of procedures. Like algorithms, it is a matter of recalling and following each of the correct steps.

Another useful skill which is often down-played is seeing what others have done and adapting minimally their example to your particular case. And yet, why not realise that this is a very useful and important technique, both in life and mathematics. I have seen some very incompetent people doing a very good job in this fashion. I have also seen some outstandingly intelligent and capable people making a complete mess of things because they wanted to begin from scratch using their own methods.

No, the ability to look up an example in a book and apply it to your particular case is a very useful skill. And learning algorithms is also very important.

Of course the educationalists are right in saying that there are other more difficult tasks and maybe we need to teach students how to do them. There is indeed a tendency for the weaker student to rely on the standard problems and the algorithms. These are an important and valuable part of mathematics, but there must be some emphasis placed on the solution of problems. The educationalists have a point. But let us not neglect the value of the mere copying of methods and the using of algorithms.

In addition, the ability to convert a problem into mathematics is also a very important skill that needs to be taught. Students should be able to deal with word problems, for instance. Mechanics seems to me an excellent subject for this skill of re-stating the problem into mathematical problems which can then be solved.

In general one would like a student to have the ability to read a mathematics text-book alone at the end of the course. Also of importance, is the ability to skim a mathematical text-book, to find out roughly what it is saying. The reason for this is that one can't spend three months reading a text-book in detail, only to discover at the end that it has nothing of value for the particular problem in hand. So I am suggesting a course in which one reads two or three books, but only in outline, not in detail, so as to get some overall understanding of the subject matter in each book.

I also think it is important to study some mathematical history. I would like to see some detailed study of a small part of history, for instance, to follow the concept of function, from Newton till today. (Useful information on the teaching of the history of mathematics is available in Burn [1997].)

I mention the content of courses because that is something that we tend to inherit on arrival at a university. Of course individual courses are usually subject to revision, but seldom is the whole structure considered. Once every five years such a structure could do with a major overall revision. This should be part of the university's routine plans.

It also seems to me that it would be appropriate for such students to also study and know the "soft mathematics" already recommended for the general student. Very often, these wider aspects, which give interest and motivation, are the ones which are neglected in the traditional courses. Even if they are mentioned, since they do not count for examinations, they are not taken seriously.

One extra point I think is worth making. I do not like courses whose main aim is to teach material which will be of value only later on in the student's development, but do not have any pay-off now. It is like building the foundations of a house, but never actually getting round to building the house. Actually it is worse, because most people have seen a house, and can understand what building the foundations will lead to, and that is not the case with mathematics. An excellent example of how to avoid a course which is basically preparatory is Herstein [1996], which features interesting and important theorems, rather than technically valuable material. Of course it is a matter of balance, and some basically preparatory material is not only acceptable, but even desirable.

One needs of course to keep an eye on what employers want when deciding what subjects are to be included in a course. For that reason I will go on to the McLone report in the next section. However, there is an important point I need to make before then. There is a great deal of detailed, conscientious, careful work needed to design a course, to decide which parts are relevant and interesting for the particular students and the needs of the course, to decide the order of topics and the sequence of problems, and without which the course will flounder. It is difficult to say something about such work in general, except that although not as highly regarded as it should be, it is of the utmost importance.

5.3 The McLone report

First McLone's table 3.9.

	Subjects named by employees
Over 70%	Matrix algebra, Methods of Calculus, Real Analysis: single variable, Ordinary differential equations, Vector Algebra.
Over 60%	Complex Analysis, Group Theory, Set Theory, Probability, Vector Analysis
Over 50%	Real Analysis: many variables, Partial Differential Equations, Statistical Methods

Fig. 5.1 Essential subjects for an undergraduate mathematics course

McLone [1973] was a report on students who had been trained as mathematicians in England and then went to work in industry. It is interesting to study it to find out what the employers and employees thought about the subjects studied. Being 25 years old it should be regarded more as a useful model for new investigations, rather than as an authoritative study, especially in view of the large changes in industry.

The above table summarises the response of the graduates to the question of what they thought were essential subjects for their work. Today I would suggest that numerical analysis, combinatorics, computing, and operational research would be other subjects that would appear in the table.

The following table is a free adaptation of tables 3.13 and 3.14 in McLone's report. The figures that appear come partly from McLone's version and partly from some guesses which I have made, and some manipulations, mainly to get a scale of 1 to 6, which makes reading the table easier. Again the responses are those of the graduates, this time to the question of which areas of ability were important in their work. McLone makes the point that with a few minor differences, the graduates' replies agree well with a similar questionnaire answered by the employers.

	Important	**Present in courses**
Problem formulation	4	3
Problem solution	4	5
Knowledge literature, techniques	3.5	3
Skill in verbal communication	4	1
Skill in written communication	3.5	1
Ingenuity in working	4	3
Ability to acquire new knowledge	4	4
Accuracy	3	3
Overall planning ability	3.5	1
Critical evaluation of work done	3	1

Fig. 5.2 Importance in work of various abilities, Graduates' opinions.
6 = max, 1 = minimum (modified table).

Problem formulation, solution, knowledge of the literature, ingenuity in working, ability to acquire new knowledge, seem to have been regarded as satisfactory. An obvious weakness is skill in verbal and written communication.

McLone goes on to say:

"There is strong evidence both from employers and graduates that the most useful areas of mathematics are contained in the numerical applications of the subject, especially statistics. Academic specialists are not in demand; a secure foundation plus a wide coverage of statistical methods is generally the most sought after combination." And "...an important aspect of applied mathematics which is emphasised by all groups is mathematical modelling, that is, the modelling of real situations in mathematical terms. This aspect of applied mathematics is vital in many areas, but is unfortunately often overlooked in university applied mathematics courses."

In urging that one should see what industry sees as the requirements for its workers I am not suggesting that we should follow their recommendations blindly, simply that we should keep them in mind.

There are several reasons why industry is not able to provide definitive answers. The first is the very nature of industry means that their perspectives have to be relatively short term. They prefer to be ahead of the rest of their competitors, but not way ahead. It is a matter of just being able to out perform the others. Trying something very much in advance of current practice can be commercially very risky.

The second reason why industry has difficulty in suggesting what their requirements are is that they are so diverse. Depending on the particular job pursued by a graduate, he or she may need to use only a fraction of the courses studied. If one knew exactly which job a student was going to take, and if that job was to remain static, there would be no difficulty in drastically reducing the content of a course. However, we have to supply an education which will fit the graduate for most jobs in the chosen profession. This means inevitably that there will be much extra that needs to be studied.

The third reason why we cannot rely solely on industry's judgement, is that often it is very hard to know just what has proven in practice to be of value. Ideas and techniques, approaches learnt in one subject, can be of great value when transferred to another, and this transferral can take place without one really realising it. So for these reasons, it is sensible to take industry's comments as helpful suggestions, and not as a prescription as to what should be done.

5.4 Modern technological methods of teaching

Multi-media, computer based teaching, distance learning, the use of email and all sorts of other techniques are upon us. Are they of any use?

My first experience of alternative methods of teaching was a series of programmes offered by the BBC in the middle 60's on surgical operations. The lens of the TV-camera made every aspect of each operation very clear and easy to follow. It was almost better than actually being present at the operation.

Similarly, there are multimedia courses for pathology. In pathology you must be able to recognise certain patterns, and this one can do only if one sees actual examples. To be able to remember them requires rote learning as well. All of this is done far better with multimedia than with a lecture.

But all of these things belong more to what I call training. We know the answers, and the student has to learn them, without question. The student's activity is directed, checked, and re-checked, automatically leading to increased knowledge. The student is not required to think in an original way.

It would be foolish to suggest that similar methods applied to mathematics will not be of the greatest importance and interest. It is too early to say if they will replace what is presently regarded as good mathematical teaching, and what effect they will have in general.

Up till now, mathematics has been a subject where a 16 year old can produce an argument, based on nothing but ingenuity of mind, with no calculator or computer, and in this way solve a problem which has stumped many a distinguished and experienced mathematician. Mind you, such a 16 year old has to be a genius. The equipment required has always been brains, pencil and paper. Is it to change now?

The solution of the 4-colour problem in which computer calculations were involved to prove remaining cases has meant that mathematics itself will never be entirely the same again, that the computer can and should play a part. Nevertheless mathematics is still for the most part the subject with minimal material. In the words of the famous saying, "In mathematics, nothing is necessary, or sufficient."

It is possible to contrast mathematics and computing in the following over-simplified fashion. In computing we do the obvious, carefully and accurately, and it used to take much too much time and require an unacceptable amount of memory. Many of us who programmed in the old days (the 70's and 80's) used all sorts of ingenious devices to reduce calculation time and to save memory. We worried about every byte. Now with the speed of the modern computer and its increase in memory, we can do many of these plodding calculations without difficulty. Mathematics on the other hand finds clever ways of avoiding the obvious and laborious solution. For instance, the method used by Gauss to sum all the natural numbers up to 99.

In the past it was impossible to do the obvious calculations because they took so long. Now in many previously intractable problems calculating is quite possible. It may then turn out that for many problems we do not need mathematics. Certain professions may no longer regard mathematics as an essential.

Thus Burton [1995], and Lange [1993] have some very interesting suggestions and projects. At times I get the impression that the idea is to avoid the teaching of mathematics, using computers to carry out the work. Thus Liddy Nevile writes in [Burton 1995]

"Looking through today's mathematics students to their future as people using mathematics, I envisage a generation of graduates working comfortably across a range of disciplines, using a range of software to perform tasks which would otherwise require a level of mathematics well beyond their immediate competence."

If Nevile is right, and people are going to be using computers to solve problems outside their understanding without the supervision of somebody who does have a complete understanding, it seems to me that we are heading towards one disaster after another. On the other hand, if the idea is to be able to solve problems with mathematics oneself, at least where calculating is relatively easy, and then to be able to solve cases where computation by hand is no longer feasible, then that is another matter altogether, and one should sensibly and thankfully use the computer's full resources.

There are other more interactive and ingenious methods of using computers in teaching. Quinney [1996] presents a multimedia course on Calculus. Whole courses on mathematics are available for free on the Web. Some books provide web sites to help their readers in various ways. But these possibilities are just the beginning.

Bonner [1995] sketches four future university learning scenarios. The fourth is very definitely many years off, so here are the first three:

The first possibility is that the university remains much the same as it is today, but is supplemented with multimedia learning possibilities, perhaps with the aid of the internet, and the use of email to obtain help.

The second possibility involves the elimination of the actual physical university, i.e. no lecture halls, perhaps not even a community of scholars. That is, courses will be accessed and studied, with examinations and tests, all dealt with electronically.

That such universities are possible is shown clearly by the success of such institutions as the Open University, centred at Milton Keynes, England. (However, the Open University places much emphasis on tutors and on summer courses.) It leads to the possibility that the university as we know it disappears completely, and is replaced by a body that designs and markets courses, which are then accessed, and paid for, on the internet. Interesting commercial possibilities arise; Oxford may for instance corner the market in Analysis courses, while Cambridge might win commercially on Algebra courses.

It can be argued that these courses would in many ways be not much different from books, and it is not at all easy to learn mathematics from books alone, but the idea is to design the courses so that they are truly interactive and in this way they may provide a considerable advantage over books.

The third possibility is that knowledge will be available when needed, and anybody can access it when it is required. The system itself could guide the learner or perhaps more accurately, the user, to the sources of knowledge that are needed. So for instance, if you had a differential equation you could not solve, you would enquire about it on the intrnet, using a search engine to get your solution.

At this stage it is premature to say too much about these possibilities. Will they take over from ordinary lectures and will the university itself disappear in its present form? It is hard to say, although the new technology and what it can achieve looks impressive. Yet many new types of technology, like TV and videos, which seemed fantastic when first introduced, have not eliminated the traditional methods of teaching. We will have to wait and see. Whatever the event, it is clear that the new technologies and techniques will have a considerable impact on mathematics education. But I for one think that the old methods will still play an important, and probably the most important role, in mathematics education. The new methods will be outstanding for cementing ideas in the memory, for drill work, for striking presentations.

5.5 Calculators, computers and lists of formulae

Many academics believe that the use of calculators has had an adverse effect on students' mathematics study. There are of course different view-points.

There are those, the modernists if you like, that on the contrary say that the skills of being able to add, multiply or divide numbers are no longer needed. We can in a fraction of the time use a calculator, so why learn something which is not necessary? Just as we do not bother to learn and practice digging holes in the ground by scratching in the soil with our bare hands, but instead use a pick-axe and spade, or even better, a mechanical digger, so we should not bother to learn how to multiply and add without a calculator.

Some of the modernists even argue that there are other advantages with using the calculator. It is so quick that one can use it to amass information and check ideas very quickly. It gives us a chance to experiment, simply because otherwise it would take so long to calculate by hand that we would never be able to do all the calculations we can now do. It gives us a chance to expand our knowledge. It is a help to mathematical understanding.

Many modernists go further. They say do not restrict yourself only to the calculator. Do not hesitate to use the graph plotting facilities of a hand-held calculator. Consider the computer as an essential part of your learning. Exploit the fundamental and easy to use spread-sheet programs.

There are others, the traditionalists if you like, who argue the opposite. They point to their experience with students who come with the mini-calculator built in to their anatomy like a third hand. The students are unable to do the simplest problem without the use of the all-knowing authority, the calculator. If the question does not involve a procedure which they have on their mini-calculator, then they are stumped.

The traditionalists hold that adding, multiplying and dividing numbers gives one a more concrete understanding of numbers. If one is shown an unfamiliar object one automatically says "Let me see that," meaning really, "Give it to me, so I can hold it and feel it." In that way one gets a better understanding of the object. So in the same way handling numbers by oneself gives one a better understanding of them. The simple $a^2—b^2 = (a—b)\ (a + b)$ is more intelligible, for instance, if one substitutes various values for a and b and subsequently calculates.

The traditionalists have another point. Arithmetical operations are algorithms. One does one step at first, and then another depending on the result of the first, and so on. If one is to learn how to use algorithms in general it is helpful if one has practised on a number of algorithms, and the simplest are these algorithms of adding, multiplying, and dividing.

Also if one wants to be able to accurately carry out algorithms and other operations, then it is useful to have experience of procedures which require care and accuracy. It is an additional reason why basic number manipulation is useful. Thus, the careful detailed arguments which are needed in drawing a graph of a rational function, involving a table of signs of the derivative, also give skills in careful reasoning. If these various methods which involve detailed and accurate arguments are removed from the students' study and are not replaced by anything else, the student is at a disadvantage.

My own feelings on this matter is that both views have some validity. I definitely feel that hands-on calculating and even drill is very important. The traditionalists are very definitely right on this point. As for calculator/computer vs. personal calculation, yes, it is useful to be able to use tools. But the main question is who is the master? Is it the machine that decides the result or is it the person? If the person must defer to the machine, and has no other way of checking the result, but just blindly types in the figures, then understanding is insufficient. So although the modernists are right to stress the value and the saving in time that rises from the use of modern computing machines, that must take second place to understanding. It is paradoxically those that can do the calculations by hand that stand to benefit most from the use of technology. They can use it to improve their understanding as well as to shorten the time required for computations.

There is another phenomenon: the rise of the formula list. This is a list which is provided at the examination. The argument is that it is silly for students to spend so much of their time learning things. After all, in ordinary life one does not hesitate to reach for a reference book on one's book-shelf. Why should students not also have this advantage during examinations? Is there any advantage in actually learning things by rote? I think there is.

A disadvantage with providing lists of formulae in the examination is that it suggests that it is not necessary to learn the basic formulae and theorems. Admittedly provided with a formulae list one can solve one-step problems. But whenever one needs a flight of imagination, whenever one must do something which does not follow directly from a formula, the list of formulae does not help. One must "know" the formulae in some sense, so that one can have that sudden flash of understanding, and connect those formulae or theorems with the present problem. When one does not know something very well, it is extremely difficult to realise that it may be of value. Some learning is a necessity.

5.6 Conclusions

Universities obviously need their own research and thoughts into what will be required in the future to decide on teaching for the next decade. As far as

I know, there is no university which takes this idea seriously. Yet this surely is what we as universities should be mainly responsible for?

The new technologies may mean far-reaching changes in our practices, and it is hard to predict what will happen at the moment. Universities are quite conservative, and so the changes may be slower than predicted; still there will be changes. Whether they will be for the better or worse remains to be seen.

6. Study Skills

Preview

Study skills need to be built into the process of education. Indeed, it could be said that the main aim of the university education is to make students independent, able to study on their own, able to read a text-book entirely on their own, and able to solve problems on their own. In the past, students came who had already learnt quite a lot on how to learn at school. They learnt to refine and improve their methods of study by means of coping with their university courses, but now we need to help them more.

As a lecturer one needs to remember how the student is going to learn so that one can arrange the lectures to facilitate learning.

6.1 Steady study

Most important is a procedure to ensure that work is carried out steadily and regularly. The student should arrange for the university's time-table to cue the work, deciding on simple procedures which must be performed. For instance, the student can decide to alway write careful notes or a summary after each class period, and to always prepare and revise class-work (cf. §6.8). In this way the work is covered automatically and methodically.

Students must learn to keep their cool, retain hope, and realise that most do not understand the material at the first attempt. Since in the past the majority of students mastered the material in the end, there is no reason why the majority of the present students shouldn't learn the material too, but it will require hard work.

Students should realise that understanding take time and that often a proper understanding will only come on the last revision, just before the examination.

6.2 Memorising is important

The student needs to memorise material. Unfortunately one of the tenets of the school training in several countries is that one should never memorise anything, since one should either understand it or look it up in a book. This is wrong. The student must memorise the material, preferably by understanding it and being able to derive it quickly, or else by using mnemonics. It helps if the lecturer can find as many of these so as to reduce the burden of memory. But the student needs to realise that rote learning has its place.

Usually it is best to understand something before committing it to memory, but it can be a useful tactic when one has failed to understand something, to learn it off by heart. Understanding can then follow subsequently. I do not recommend learning vast quantities of material in this way, but when one cannot make further progress in studying something, one can profitably learn the item off by heart. It is then available for use, and furthermore, a subsequent attempt to understand the new idea is then often successful. This is better than the other alternative, of simply giving up when one just can not understand something.

It is helpful for the student to write in a note-book the main definitions, examples, and theorems in summary form. The student benefits enormously by producing such a summary, day by day, and learning it, steadily, week by week.

At the end of the course, the student should indicate in a diagram how each item is related to the other, a lengthy but worthwhile task. Instead of a diagram or else in preparation of such a diagram one can produce a dependence table, like the example below. The example considers only a fragment of a course in linear algebra, but the idea is to tackle the whole course in this fashion.

No.	Classification	Statement	Depends on
1	Definition	Linearly independent	
2	Definition	Spanned by a set of vectors	
3	Lemma	In a space spanned by n vectors, $n + 1$ vectors are dependent.	1 & 2
4	Definition	Dimension	3

Fig. 6.1 Dependence table

One should not be too dogmatic, because for many students these methods may not be of value. Still, the more methods one suggests, the likelier it is that students will find something suitable. The point is that students must find their own methods, but can very well begin by considering your suggestions.

One such suggestion is the so-called "Cornell method." Students are advised to divide the pages of their note-books into a narrow column on the left, and a wide column on the right. The right-hand column is for detailed notes, the left, which is headed recall, contains one or two key-words which summarises the corresponding material in the right-hand column. When studying one's notes, one covers up the right-hand column, and uses the left-hand column to prompt one into recalling the material in the right-hand column.

The student should be aware of the advantages of discussing the material with fellow students, and the importance of revision.

6.3 Reading a book

The student should be taught how to read a book. One should learn that one can and should skip ahead, one should try to find out which are the important results and concepts, and one should ask a number of questions.

The student must learn the art of asking questions. Too often questions outside a very narrow range are discouraged. Questions like "What is the point of this?" are sensible and obvious questions. Often we do not know the answer, and so instead of admitting it and saying we will try and find an explanation later, we tell the student to shut up and wait till we have got through all the material, promising that then understanding will come. But usually it is possible to give an explanation, perhaps not the best, but at least some sort of explanation. Later on, it is true, that when all the main ideas are understood, it is easier to give a better justification. But it must be a spiritless and cowed student who will take "keep quiet and you will find out later," for an answer.

There are a number of obvious points which need to be taught. Thus it should be explained that in most text-books one line may follow from another, but often intermediate steps are left out, and that is why one needs pen and paper, to try and provide the missing steps. It should be explained that the book will often assume the student is able to fill in the missing steps, particular if something similar has been explained before. It helps to illustrate this. The following is an example of how this could be done:

In Haggarty [1992] we read on page 67 the following:

"Since $x^2 = 2$, some simple algebra gives the rather strange equation $x = (x^2 + 2)/2x$."

In order to see that this holds, one must do some work oneself. Thus taking the right-hand side and substituting 2 for x^2, one sees that $(x^2 + 2)/2x = (2 + 2)/2x = 2/x = x^2/x = x$, i.e. the left-hand side. That is, Haggarty is right.

Students who simply stared at the original sentence without trying to write down something might have spent a long time without getting any further. But if one attempts to do the calculations oneself it becomes much easier. It is a good idea to give the students several examples like the one above.

Point out that if the students are trying to follow an argument and there is a large complicated algebraic calculation which they cannot follow, they should skip it. Instead they should try to follow the essential details of the argument, in the confidence that with a little struggle they can return and work out the details of the calculation.

In following a proof or argument which is complicated, the student should try to split it into blocks of argument. For instance, to prove this result we need to apply theorem A. The first part of our argument is to show the conditions of theorem A apply. In order to do this we will need to use Theorem C, etc.

There are a number of useful techniques for study which are worth mentioning. I refer to them as SQERPSR^2 (pronounced squerps r squared), The Jig-saw Puzzle Method, and The City Explorer.

The method briefly described as SQ3R for reading English texts is the inspiration for the first method. In this method one begins by skimming through the book, then asking questions, and then reading carefully, and then reciting and finally reviewing. This is a sensible way of reading a book. The corresponding way of reading a mathematics book can be described by SQERPSR^2.

Again S stands for survey or skim. Thus one should note the organisation of the chapter or book one is studying, the main theorems, the main concepts, and try to get some idea of the methods of proof. Q stands for questions. E stands for doing a number of worked examples. Then R for a careful reading, followed by P for problems. S stands for writing a summary, as suggested before, of the main results, definitions and formulae. The first R in R^2 stands for recite, i.e. memorise, while the second R for review or revise.

It is important not to think of this as a magic formula which is the passport to success. It should not be applied blindly, for there are certainly times when it is inappropriate. But at least it gives the student two advantages. The first is that the student realises that it is totally insufficient just to read the text-book, that a large number of steps are required to master the work. For the second, it gives a procedure for coping with the work.

I think it is worthwhile doing an example of the procedure with the students. Thus for instance, I would take a book that we were studying, e.g. if we look at Chapter 3 of James [1993].

S stands for skim. This chapter is entitled Group Representations. There is a helpful introduction, and we get the vague idea that a representation has something to do with a homomorphism and matrices. The first section seems to give the idea of a representation, indeed, it contains the definition, and then there seem to be some 2x2 matrices.

The next section talks of equivalent representations, there seem to be conjugates by a fixed matrix T. Then there are kernels of representations, this must have something to do with kernels of homomorphisms. Finally there is a useful summary.

Q is for questions. So the obvious questions would be what are these various ideas, what is a group representation and what has it got to do with matrices? What has the lot to do with Chapter 2, which was about vector spaces and linear transformations? I can see it has quite a lot to do with Chapter 1, which introduced the idea of a group and spoke about group homomorphisms. Then too a glance at Chapter 4, which is about FG-modules, is not too helpful; it is difficult to know how this relates to Chapter 3, so perhaps there is a need to wait till later before considering this aspect.

E is for working through the examples. So now carefully work through the examples in 3.2. In order to understand them one is forced to read Definition 3.1. Then one would now check Example 3.4, and then 3.8.

R is for a careful reading. At this stage the text should be carefully read, starting at the beginning. Again, the student should remember that one should read with pen and paper, and break off the reading to derive the missing lines.

P is for problems. Having done this, it is time to try the problems on page 36.

S is for summary. Now is the time to go over the text again, writing a summary of the definitions and theorems.

R is for recite. Write down, without referring to the text, all the statements of theorems and definitions, and subsequently check whether they have been recalled correctly.

R is for review. A helpful review is in fact given by the introduction and the summary, both of which can be read very carefully now. It is also worth quickly reading through all the definitions and theorems.

Actually, I doubt whether I or anybody else would have gone through all these steps in practice, which underlines that one should not follow a method slavishly. Still it does illustrate the principles.

By going through this procedure, the student is preparing the ground work for understanding, even if the material does not really make much sense at first sight.

There is another useful method of learning, which is the very opposite of the method above, and which I call The Jig-saw Puzzle Method. It corresponds to the method of completing jig-saw puzzles, the sort where one is not given the final picture but just the pieces. One assembles little parts of the jig-saw that make sense by themselves, and then assembles the completed parts together to form a whole. One gets little groups of insights and only then builds them up to make a whole. One method is therefore a top-down type of learning, the other a bottom-up method of learning. Both $SQERPSR^2$ and the Jigsaw-puzzle method are helpful methods of learning. The reader may not like this, because it sounds like a contradiction. But people learn in different ways at different times, and sometimes one method is better than another.

If one comes to a strange city one normally starts off with a fixed point, say the railway station, or one's hotel. Everything else is then related to the fixed point, and other points are then related to these new fixed points and so on. This is also a useful stratagem in learning, which I call the City Explorer. Start off with something you know and understand well and then relate new ideas and results to the well known idea. This is also similar to the historical method of learning, where one repeats the steps taken in the past (at least to some extent) and in this way obtains a better understanding. This is beginning with something you know and progressing one step at time, relating each step to its predecessor.

Bearing in mind these strategies of learning, one can make one or two comments about styles of lecturing and also of books.

Some lectures and books use a very terse style which require one to begin at the beginning and master each step perfectly before proceeding. Typically unnecessary words and explanations are avoided and one relies on a precise and concise style with brief notation used without further reminder. The same books and lecturers often choose to assemble the precise material needed, and faultlessly and without explaining any connections, derive the major results in the most general form.

It is clear that such methods effectively prevent one using the jig-saw puzzle method or the city explorer or the historical approach. I am not saying these books are not good; often they are models of efficiency and clear logic. But personally I appreciate such clinical perfection when I have to some extent mastered the basis of the material, not when I am learning it for the first time.

6.4 Solving problems

For the solving of problems, students need to be taught the simplest of methods. First one must understand the problem. One should do some calculations which enable one to get a feel for the problem. One can draw a graph or a diagram. Look for a theorem or definition or example in the text-book which looks similar. While one is doing this, one can try to learn these relevant theorems, definitions and examples. If one does that, at least one has accomplished something, even if one has not solved the problem.

Show the students how if one is trying to prove that P implies Q, that one can both work forward from P and backward from Q, trying to find a point where the link between them is closer. It helps to express this as a maxim: "Move the beginning to the end and the end to the beginning." (See Appendix G for maxims.)

The students should also learn the technique of guessing. This technique is almost always valuable, even if the initial guess is stupid. For instance, suppose one is asked to solve a system of linear equations in 3 unknowns. One guesses a solution, say 1. Then one asks oneself how one would check that solution. Obviously by substituting. Then one realises that one needs a solution with three numbers. So one tries 1, 2, 3. This is a wild guess, but as a consequence one substitutes in the equations and in this way obtains an understanding of what the problem is about. Indeed, a guess is nearly always helpful, but with one proviso: it must be followed by a check. As a maxim, I put it in the form: "Do the mathematical two-step, guess-check, guess-check." Thus the guess is always followed by the question "How can I check whether my guess is correct?"

Not only is it desirable to check guesses, it is equally important to check solutions, even if one has used a standard method. When one has used a computer it is particularly important. Most students do check their answers against those of their friends, or check the crib if there is one. It is not such checking that I am urging, but some independent way of finding out whether the argument is right or not. Finding such methods can be difficult and requires ingenuity.

The easiest way to teach the student these methods is to actually carry them out in classes.

6.5 Polya's ideas

For students who are going to specialise in mathematics one should, over a period of time, introduce the more interesting ideas of Polya [1965, 1977, 1985]. But these methods must be distributed over the courses gradually, and illustrated several times by means of examples that you solve.

An important principle that Polya enunciated was that having completed a problem or proof, one should examine it again and check whether one could understand it better or whether there was anything else one could learn from it.

These ideas are obvious, even trivial. And yet they will not sink in if they are not continually repeated.

I would urge the reader to read Polya's interesting books, but I will caution the reader to note that not everybody will find Polya's methods satisfactory. People think differently.

I remember for instance two friends of mine discussing a chess game they had just played. The first person explained his thinking brilliantly, logic predominated, and where necessary he explained how he had thought five or six moves in advance, and in detail. He also explained his strategic thinking, and was very clear and logical. The other player was much vaguer. His explanations were fuzzy. For instance, he decided against a move, since the consequences looked dodgy. He was afraid of certain things but he could not really explain why. I much preferred the first player's explanations. And yet it was the second player who usually won when the two played. One must be pragmatic, the most satisfying and logical approach may not be the best in practice.

In his book "How to Solve It", Polya subdivides solving problems into four steps, each of which is facilitated by suitable prompts. Thus,

First Step: Understand the problem
A typical prompt question would be "What is the unknown?"
Second Step: Devise a plan
The sort of question that Polya would ask here is "Do you know of a related problem?"
Third Step: Carry out the plan
Fourth Step: Look back
A useful question would be "Can you use the result, or the method, for some other problem?"

This then is a very brief indication of Polya's formulations.

6.6 The student's most common fault

Most students work hard, study examples and problems, and get to the stage where they can follow most of the arguments and understand their notes and the text of their book. They can even understand the solution of typical exam problems. At this stage they tend to relax. But they have not learnt or understood the material well enough. They are near to being able to answer the examination questions but not quite there. Consequently the examination proves too difficult for them, they make simple errors because they have not quite understood something or they have remembered something wrongly.

The point I am making is that over learning is required. I try to get this idea across by asking the students to consider how they would prepare for a 100 meter swimming race. Would they be content to get to the stage where they could swim 100 meters and leave it at that? No, of course not. Similarly they must practice, revise and practice their mathematics again and again. If they keep this in mind right from the beginning they are in a better position to organise time properly.

6.7 The problem of definitions

Definitions of course must be learnt, because they are more than just explanations of a word, they tell you precisely what you need to do when you need to prove something.

Students must be taught that definitions in mathematics are not like definitions in a dictionary, but they are like invoices which need to be checked. This idea needs to be re-iterated frequently.

They must also learn that defintions are often wide-ranging generalisations of concepts, that what we have done is to take the very essence of a concept and state it in an abstract and useful form. It does help to try and show why the name is what it is, and to show a series of less general concepts which finally lead to the new concept.

Students should realise that a good idea is often pushed to its extremes and generalised, so that an awkward and apparently meaningless definition has been obtained by taking a good intuitive idea and seeing how it can be generalised, just for the sake of generalisation.

Wherever possible, it is sensible to give not only the correct definition, but also an intuitive definition, which is not there to prove anything, but rather to illustrate ideas.

This idea that there are two definitions, the correct one and an intuitive one, is useful to get across. One uses the intuitive one to re-assure oneself, perhaps to get the idea of how to understand or prove something, but in the end, when one wants to prove something properly, one uses the proper definition. Just as the proper definition requires an example or two to embed the idea in the mind of the student, so does the intuitive idea require a couple of examples. It is crucial to emphasise that use of the intuitive definition in an argument does not constitute a correct proof; it merely adds confidence and intuitive understanding. (See also §11.8.)

It is standard practice (and essential) when introducing a new definition to give a number of examples and counter-examples. Each student, should as a matter of course, take one of these examples, and learn it together with the definition. This makes it easier to remember the definition. This is not surprising, because it makes the definition meaningful.

To repeat, I am making a point other than that we should illustrate each definition with the help of an example. The point I am making is that we should encourage the student to learn not only the definition but also to learn the example, and that learning both is equally important. This is as important as learning not only a new verb in a foreign language, but also how one conjugates the verb.

For instance, when I introduce eigenvectors and eigenvalues, I take as my standard example the 2x2 matrix A with 1 in every entry, with eigenvalues 0 and 2 and eigenvectors $(1,-1)^T$ and $(1,1)^T$. The matrix and eigenvectors and eigenvalues are easily committed to mind, and I urge the students to learn both the definition and this example.

Although this procedure sounds as if it is giving the student more to learn, it does in practice mean it is easier to remember the definitions. The same procedure should be used with theorems, that is one should not only state the theorem, one should also provide some standard example which is constantly recalled whenever the theorem is recalled. Students will find this a useful method.

6.8 How to use lectures and tutorials

Lectures and tutorials should be regarded as resources, which students can use as they please, but the more work they put in the more they benefit. I compare lectures to a sandwich. The bottom layer of bread is the student's preparation, without it, the lecture, which is the filling, falls to the floor in a mess. The top layer of bread is the student's work after the lecture, without it the filling cannnot be retained.

I find it useful to make the following suggestions.

Suggested working practices:

General:

- Attend all lectures and tutorials without fail. This is vital, because it ensures that you will in a steady and systematic way work through the course.

Before lectures:

- Prepare for each lecture. You will find that even a little preparation makes it much easier to follow lectures. For each class hour I suggest ½ hour of preparation.
- Learn the previous definitions, formulae and statements of theorems. Remember the maxims "Honour your theorems and definitions for the rest of your days," and "A fool and his theorems are soon parted."
- If you have a suitable book, read ahead. For instance, you can list the new definitions, formulae and theorems. You need not understand them all, but you can among other things, try to see how they relate to the previous material and how they relate to one another. You can also see whether there are some ideas, examples or problems that you can understand.

After lectures:

- Work through what was covered in the lectures carefully, as soon as possible. For each class hour spend at least an hour afterwards.
- See whether you can state the results without using the notes or the book.
- Write a summary of the definitions and statements of theorems, including examples.
- Work through a number of problems.

Once a week:

- Revise the material of previous weeks. Devote at least an hour per week for this. Remember the maxim "A revision in time saves nine."

Before the examinations:

- Special revision is obviously needed.

The above are suggestions, and students will obviously modify them, the important thing is to get the students to realise that steady, systematic work is required. I must caution the reader against assuming that students will enthusiastically embrace these suggestions or any others for that matter. Most people are conservative, and unwilling to change their habits. Students may not be sufficiently committed to the course. Sometimes they will be willing to try the suggestions, but find they do not help them, or else find they do not have sufficient time. However, making the suggestions right at the beginning of the course gives the student a frame-work for organising study.

6.9 Techniques of final revision

Students can also do with some tips for how to revise for their final examinations. One method is to divide the course into quarters and begin with the last quarter, going backwards to the earlier quarters if needed to comprehend the material. Then one studies the third quarter of the course and so on. There are several advantages with this method. The first is that often the earlier parts of a course are there in order to prepare the way for the latter parts of the course, and one builds up a better understanding by beginning to study what one would like to establish. The second advantage is tactical: often in examinations the latter part of the course is over represented. The third is that studying in a different order makes the study more interesting.

Another method is to collect, while studying the course, a series of very easy problems which illustrate every idea and topic. This means that these problems can be done very quickly, and this in itself provides a good review of the material. Doing old exam papers is also a good idea. But one should do at least one of these as if it were a proper examination, i.e. within the set time and without the use of aids like text-books or notes. Revising and learning the summary of the definitions and statements of theorems is also a useful method.

Many find it useful to work through their notes. Working through the text-book is less effective, since one saves time if one has condensed and summarised the material in note form. Often it is sensible to use several or even all of these methods. Finally a very quick run through of the whole course the day before is a useful additional measure.

6.10 Summary

It is helpful to formulate study skills explicitly and to give examples in lectures and or class-work. Since these methods will help the students to study independently they should not be regarded as incidental but as a valuable part of the course, requiring and meriting considerable time and effort. I have found that stating these ideas once or twice is ineffective; they need to be repeated and re-explained if students are to master them. Of course nagging is never pleasant, but one needs to be close to nagging the students if one is to get the importance of the ideas across. One can vary the nagging by quoting maxims, such as those in Appendix G.

7. Rules of Teaching

Preview

How one teaches must, one would think, be strongly influenced by theories of learning. To some extent this is true, and theories of learning are discussed briefly in this chapter. But the theories of learning do not provide a sufficiently clear guide to help teaching, so instead of using any one of these theories, I have chosen a pragmatic approach and give a number of fundamental rules of teaching which are helpful in analysing and improving teaching.

7.1 Theories of learning

There are a number of different theories of learning, which give a way to interpret the problems faced by teachers, and again and again cases seem to be easily interpreted and understood with these ideas. People who have accepted these theories find confirmations of the ideas cropping up repeatedly. They very much value the insights they obtain in this way.

These theories vary from the now ridiculed simple tape-recorder model (the teacher speaks and the student automatically absorbs) to, for example, constructivism (the student must need to create the material anew in his or her own peculiar way, unique to him or her), which is now favoured by most.

A useful, simplified account of these various theories of learning is provided by Phillips and Soltis in Phillips [1991], who among other things, discuss: Plato's recollection theory (in which we remember things from a previous experience). Locke's "blank tablet theory", in which we come almost like a clean slate and listening to the lessons corresponds to the slate being written on. Behaviourism (in which reward and punishment play an important part). The Gestalt approach (in which ideas are mastered as a whole, rather than assembled as small bits). Dewey's theories, in which emphasis is placed on learning through doing and experiencing as we solve problems, leading to the idea that we must be active rather than passive learners. Piaget's theories, based on his observations of development of children. Then there are the social theories of learning, as exemplified by Dewey, Vygotsky and Badura. Finally, they discuss the Cognitive Science approach, which seems to call upon our experience with the new machines such as computers.

In my mind, however, although I feel most of these theories seem to embody some of the truth, I am tempted to say that they must in some sense be combined, for none of them contains the full essence of learning.

The support for the theories of learning is to a large extent introspective (one simply checks in one's own mind whether one thinks that way) and it is also based on observation, for instance of small children learning, and through interviewing people as they learn.

Thus these theories are not easy to confirm in any objective way. It is difficult to see how they could be refuted. Every apparent failure can be explained with the theory. (In that sense, they are what Karl Popper called pseudoscience.)

Being a firm believer in one or other of these theories does lead to a different way of teaching. There seem to be no universal agreement that better teaching has arisen from these methods. Often this is because of the difficulty of assessing new methods of teaching, for instance often the new course has different aims, and different assessment methods (oral exam instead of a written for instance), or a different syllabus, or better results have been obtained at the cost of students and staff working much harder. (Chapter 13 considers assessment in more detail.)

Almost all these theories, starting from the work of Dewey, seem to lead among other results to the key words and phrases of "active students," "student-centred learning" and so on, ideas which are worth keeping in mind.

In the words of the ancient Chinese proverb (quoted in Griffiths [1974]):

I hear and I forget
I see and I remember
I do and I know.

It is also worthwhile remembering ideas like competition, co-operation, fear of failure, pleasure in success, the pleasure of learning, which are neglected nowadays.

Perhaps the major distinction in practice is whether one should opt for an exposition style of teaching, a Socratic discussion style, or a "discover-it-yourself" style of teaching, or some combination of these methods

To discover something oneself is obviously a good way of learning things. Students should always be encouraged to have a go at a problem themselves, before they see the solution. However, it takes a very long time to discover things oneself. If one has to study everything this way, one can not get very far.

One weakness of some of the alternative methods of teaching is they have neglected the in-built human capacity to watch somebody else at work and learn how to master the skill from the watching and then practising. I really think that teacher-centred learning is very important, and that most of the educational theories are deficient in not recognising the value of such learning. Discover-it-yourself is very good in many ways, but it should not lead to neglect of teacher-centred learning.

Of course teacher-centred learning is student-centred in the sense that everything discussed and dealt with is directed towards the students, and their needs and problems continually guide the teacher.

Most teachers, no matter their view-point, seem to agree on some common conclusions. Among these I emphasise the following fundamental rules of teaching. It is these together with the use of the attainment levels discussed in §2.5 which I claim provide a practical way to detect the strengths and weaknesses in a given method of teaching, and to suggest ways of improvement.

7.2 The fundamental rules of teaching

In this section I present ten rules of teaching. They are clearly interconnected, but enunciating them separately provides a useful way to discuss teaching.

Rule 1. Teach at the right level.

That is, ensure that the beginning level of the course is at roughly the same level as the students' knowledge, and the course maintains a level which is compatible with the students' ability. This is fundamental, because get it right and the battle of teaching is half won. Get it wrong, and no matter what you do, the course will be a failure. This is because mathematics is linear, D is likely to depend on a proper understanding of A, B, C. Even if A, B and C, are understood, most people still find it hard to understand D. So without A, B and C there is no chance. You may give the most brilliant, the most original, the most interesting lecture of all time, but if the students do not know the prerequisites, your lecture will be boring, tedious and incomprehensible. I know of nothing more boring that a mathematics lecture where one does not understand the basis. I know of no other subject where it is possible to be so consistently boring without even trying. So make sure that your students know the material you expect them to know at the beginning.

This is more easily said than done. Often a revision course is tacked on before the main course begins to make up for the known omissions in the knowledge of many students. As a revision course, it may be fine, but since the student has often not studied the material, it is not revision that is required, but a whole course.

The first fundamental rule of teaching requires for its implementation:

Rule 2. Make the class as uniform as possible.

By this I mean firstly that you should try not to combine too many different groups of students. It is not sensible for instance to teach economics students and computing students and physics students in one group. Secondly try to ensure that all the students have roughly the same pre-knowledge and the same ability.

An enormous problem is caused by the wide variety of abilities and knowledge that the students have. In England the school studies end with a final examination organised by a number of boards. In the report "Tackling the Mathematics Problem" LMS [1993] topics in the former core are listed together with the information of whether they are included in the various boards. The list shows that often a topic is missing from at least one board. This of course makes it difficult to provide a course which is intelligible to all.

It helps an individual lecturer to check the contents of the school courses by looking through a text-book at that level, or else checking through the syllabi, as suggested in §1.3. Having done that, I think it is worthwhile to provide an entrance test. This seems the simplest and best procedure.

To quote Kemeny [1964]: "Our basic assumption is that no one or two programs can fill the needs of all students of mathematics. We have therefore designed four programs."

And,

"I realise that ability grouping is often a controversial topic, but whether your own institution does this in any other field or not, I feel it is a must in mathematics."

He goes on to say: "Next there is the question of whether it is fair to separate out a relatively small number of really able students. This I want to answer categorically, by saying both the students separated out, and the students not separated out would profit."

I agree with these comments. However, several courses clearly require extra staff. This must be set against the fact that where students have a limited choice of course, inevitably we have the tragedy of students struggling and failing to cope with a course that is beyond their intellectual level, while at the same time brilliant students learn too little.

The choosing of the content of the course is often not in the hands of the lecturer. The syllabus may have been laid out by others, and the subjects included may be essential for subsequent course. This means it is often impossible for a lecturer to arrange to give the course at a slower and more appropriate pace for his students.

This re-iterates the point that the problems of teaching are not capable of being solved by the individual lecturer alone.

Rule 3. Remember to deal with the obvious.

Often one is so used to certain ideas that one does not even know how to explain them, and they seem so obvious, that one does not realise that to understand something one requires these ideas. For instance, often when I ask my students to prove that a certain subset is a subspace, I realise after a period of mutual surprise (I that they do not seem able to solve the obvious problem and they in complete bafflement) that what is missing in their comprehension is their understanding of the specification of the subset. They may not even realise it is a set! But if one misses the obvious the instruction will fail to be at the right level.

Something which seems obvious to most mathematicians and is therefore often overlooked is the ability to use language correctly. Students find difficulty in handling "for all" and "there exists" and "If P then Q" etc. In other words, elementary logic. These things should be explained, preferably in a number of different courses. Examples of the incorrect use of language should also be given; these are difficult to think of oneself but can be readily found whenever one marks examination scripts. Give the students practice, constructive advice and criticism on how to write mathematics.

Rule 4. Ensure your students are active.

The students must work with the material. Indeed, mathematics more than any other subject, is one where you must take an active part.

It is for this reason that some castigate lectures. They say that lectures are not active, all the student does is to copy notes, mindlessly without thinking. Some even go as far as saying that lectures are useless.

Well, lectures need not be inactive. Major problems will always arise if the students are out of their depth, and then lectures are not very effective. This is more a problem of the lectures being at the wrong level, or the students not using the lectures properly, rather than the lecture system itself being unsatisfactory.

Indeed, the student can and should be active when listening to lectures. (Also see §6.8) Firstly he or she will be going over the topic for the first time, getting a rough idea of how things work. Then there will be a number of points that he or she understands and remembers. Then there will be certain things the student does not understand, and will note, that these should be tackled later. There will be questions the student will ask, either of him/herself or of the lecturer at the time, or later on when there is a chance for discussion. Some ideas stick, even without the student trying. Later on when working over the material again, the student will see how they fit in.

But it is nonsensical to assume that the students will learn the material just from listening to the lectures. The lectures are just a start. They must then be re-studied in detail, and the student must work through many examples and solve many problems.

In fact, the value of lectures aside from the fact that we see a mathematician working at his skill, is that even though one does not fully comprehend what is being taught, one is going through many of the steps described in $SQERPSR^2$ (explained in §6.3).

Thus one skims through the material (even though the lecture may be complete in detail, one is not really able to follow the material properly, and so this is equivalent to skimming.) Secondly the lecturer raises a number of questions, and the student notices gaps and misunderstandings which raise further questions. Finally the lecturer probably does a few examples. The student then has to go on to the rest of the process, beginning by first reading the material again with great care.

To minimise the value of lectures is to ignore the importance of teacher-centred instruction, which as mentioned before, is a most natural way of learning, starting from childhood.

It is easier and more fun to listen to a lecture than just to read it. I note that there are some TV programs which could just as well be presented as a written page. Cooking for instance. Why not simply list the ingredients, and the instructions?

Instead, we see a cook telling us how to do the cooking, and carrying it out. We still need the recipe later, since it is unlikely that we can remember it verbatim, but it would be a most boring program with only the list of ingredients! It is most valuable to see how the cook goes about his work. Often a demonstration gets the ideas across effortlessly, whereas reading would be extremely difficult.

There are however difficulties with lectures which one should not ignore. If the lecture is given at too fast a pace the student may be reduced to simply copying what is on the blackboard, rather than thinking about the material. Deciphering the lecturer's handwriting may require most of the student's attention. The student may in the end simply copy the notes and not think.

There is the danger, in courses not based on a book, that the student sees the purpose of attending the lecture is merely to write down the information. To overcome the problems of mechanically copying notes, it is possible to give the students skeleton notes, so that they do not need to write down everything. The idea is that then one can give the student ample time both to write down and to think about the most important points. Some lecturers have tried to give the students complete notes. I have no proper reports about this method, but the little I have heard indicates that it is not effective, in that students find the subsequent lectures boring.

Certainly excessively long and too many lectures, and instruction based only on lectures, is not going to be successful. The student needs to ask questions, to get an overview, to go through the details, to do problems, to discuss, to revise, to summarise, to relate the new material to the old.

There seems to be a tradition in Sweden to have lectures of 2 or 4 hours duration. (In Lund students in the first year attended a problem class for two hours, and then went to attend a two hour lecture.) This is madness. It is true that there is a break of a quarter of an hour every hour, but listening to a lecture for two hours or more is ridiculous. A glazed look comes over the students, and you can wave a hand in front of their eyes without them blinking. A glazed look often comes over the lecturer's eyes as well.

Long lectures are not satisfactory, but shorter ones often are. Indeed, the student who has not attended lectures is at a disadvantage in that not having had the opportunity to learn how get the best from lectures. So often ideas—especially new ones—are presented in the form of lectures. The ability to absorb information from a lecture, to take sensible notes, and to re-study the notes, is a valuable one.

Rule 5. Make demands.

You will seldom get more from a student than you demand. So demand a lot. That is the only way to get a high standard. Just as an athletics coach will not be satisfied with athletes attaining a certain level of speed when they can do better, so one needs to be firm, tough and demanding to get the best from one's students.

One of my colleagues said there was a universal constant, K, very much smaller than one, such that the students learn KX whenever they are taught an amount X. So if they were taught more, they would automatically learn more. This overstates the case a little. All the same, one has to set a fast pace, and give difficult assignments. This must be done with some discretion. It is easy to give tasks which prove impossible for the students one is dealing with, and this is counter-effective. In this, as in everything else in teaching, balance is required.

The point is that nobody can continue if they meet too many failures. A useful point to consider is the order in which you set students problems. On the whole the student expects the easier problems to come first. Having tried and failed the first three problems on the list students can give up in despair. Yet the next ten can be well within their grasp. So either change the order or give some indication of which problems are likely to prove challenging.

Rule 6. Encourage and give students opportunities to ask questions.

Students must be able to ask questions of both their teachers and their fellow students. In McLone [1973] we read "The criticism most often aired was the almost complete absence in many university mathematics departments of any evidence of the "meeting of minds" between tutor and student which is particularly held to be valuable in a university education."

This emphasises the importance of tutorials and discussion groups which will enable the student to speak both to fellow students and to the teacher.

In Lund this was supposed to be taken care of in the following way:

(1) Twenty or so students were assigned to a weekly problems class, where the idea was for the students to show their solutions to the class. If they could not give a satisfactory solution, the tutor would do so.

(2) Most days the students were given a large room to work in for much of the afternoon. The lecturer would come for part of that afternoon to help with any problems.

(3) After passing a written examination, the students were given an oral examination. The oral examination has been lyrically described by Professor Lars Gårding as a "never to be forgotten meeting of a student with a real-life mathematician." The students I met did not seem to view their oral examination in quite such a positive way, but there is no doubt that it did add immeasurably to the students' understanding of mathematics. I have more to say about this aspect of oral examination in §8.2 below.

If in addition to these measures, we can also include an individual project for the students, I feel we will be a long way to meeting the students need for the "meeting of minds." Once again we must ask ourselves, is it practicable to provide all these desirable items, or are they too expensive? After all, everything costs.

Despite the very well organised system at Lund, the students did not prosper as much as one would have hoped. This was mainly due to the fact that (1) did not work well for most of the students. Usually the graduate student in charge of the tutorial groups reported grimly to me that "Again it was a solo performance," in other words, the only person to contribute to the solutions was he himself. I am convinced that the reason for this was that many of the students were out of their depth, in other words, the first fundamental rule of learning was not observed. It is difficult to fault the way the system is organised otherwise.

Rule 7. Motivate.

If you do not put a considerable effort into finding ways of motivating your courses the students will struggle to keep awake. I would even express it more vigorously, "Motivate or give up." Courses at university level are and should be difficult and demanding. There are long hours of apparently unsuccessful study. Without a very strong motive students can not overcome the boredom and the frustration of the work.

Yet motivation especially in the beginning is not often stressed. There are many courses which are given in exactly the same way to completely different groups of people. Thus building engineers and physicists, electrical engineers and aeronautical engineers, biologists and business studies students, will all be given exactly the same course on calculus. How can these various students be motivated simultaneously? It is clearly impossible.

However, if you have a fairly uniform group of students, and if you take the trouble to understand their interests and what they feel are significant ideas, you can tailor your course to appeal to them, and give them interesting examples relevant to their field of study.

It can also be of value to put in the occasional application to something of general interest. I like to put in Olber's paradox when discussing generalised integrals for instance.

I have always noticed that it is the balance of the class that either lifts up the class or drags it down. If the majority are enjoying the course and benefiting from it, they pull the others up. If the majority are finding it a drag, they pull down even those who could enjoy the course.

Rule 8. Make lectures interesting.

Very few lecturers consider how to make their lectures interesting. So much material must be covered and covered thoroughly and systematically that it is very easy to be reduced to something analogous to painting a wall. One begins at one corner and slowly and thoroughly covers each part of the wall—there are no moments of excitement (except when the job is over).

But lectures should be exciting and enjoyable. Each lecture should deal with a whole topic, and be organised with a beginning, a middle, and an end. If this is not possible, one could perhaps organise two or three lectures together so that they form a whole. And just as a good serial or soap opera will tantalisingly end on a high note where one is left waiting eagerly for the next episode, so too should lectures end with the promise of something interesting next lecture.

Rule 9. Treat students as people.

Students are human beings, with problems, anxieties, difficulties. They are driven by fear and by pleasure. They need encouragement. Make sure the students know you are on their side. They need tasks which are do-able, and which when solved give a feeling of achievement. They need a positive word from one of the lecturers now and again.

It is expensive in time and effort, but it is worth assigning each student to a member of staff, to meet regularly every week or two, to establish a personal rapport. Simply listening to a student can be extremely helpful. It may very well be that nobody the student knows has any understanding of the course and difficulties involved in studying it, and your concern and interest can provide strong motivation.

It is sensible at least once during a course, for the lecturer to talk to each student individually. In this respect one can learn from medical doctors (or at least some of them), who have the knack in a few words of leaving a feeling of trust and confidence in their patients.

As a mathematician it is easy to fall into the trap of being contemptuous of the weaker students. However, these may very well have been more virtuous than the others. They may have spent many long and fruitless hours trying to learn the material, while their smarter fellow students have managed to absorb the material merely by glancing at it. So do not underestimate the amount of work the weaker students may have done, nor their commitment. They deserve and need encouragement.

Rule 10. Lecturers, keep on learning.
The teacher must continue to be a student. Only by studying further will one have firstly the humility required, realising whenever struggling to solve a problem, that the student has similar difficulties. Only in this way will the interest and love of mathematics be retained. Only in this way will the teacher have a wide enough knowledge of the subject to guide the student.

This has obvious consequences in the overall planning of the departments. Lecturers must be given time and encouragement to study, and in particular must not be overburdened with teaching and administration.

7.3 Seminars

One place to see the need for these rules of teaching is in research seminars. It is not often that these rules of teaching can be practically fulfilled here, and many seminars fail to give much to the audience.

The audience usually consists of a mixture of mathematicians, with skills and expertise in various specialised subjects. That the subject matter should be at the right level in accordance with the first fundamental rule is obviously almost impossible for the lecturer to organise. Generally the lecturer makes an effort. All the definitions needed for understanding are given very concisely without repetition. For instance, I once attended a lecture on error-correcting codes. The lecturer gave the definition of an n-block error correcting code on an alphabet A as being any subset of A^n. This is technically correct, but valueless for those who already know what an error-correcting code is, and hopeless for those that do not know. Giving the formal definition alone is not good enough, one must give the spirit, which is even more important than the correct definition.

Often there is little effort to motivate the seminar, in accordance with the seventh rule of teaching. The lecturer, who has been dealing with the subject for so long that its interest is obvious, forgets to motivate it. For the speaker the main point of interest may be how to manage to get by a cul-de-sac by squeezing past a difficulty using a particular trick. But for most of the audience that is not what they are most interested in to begin with.

Then the eighth fundamental rule of teaching, to ensure that the lecture is interesting, giving it a form, a structure, a beginning and an end, this is often ignored. Indeed it is not uncommon for lecturers to completely underestimate the time required, so that their talk ends with only half the material covered and in some confusion. The lecturer has failed to be prepare adequately, feeling that no preparation was necessary, having worked on the subject matter day and night for many years. And that is the trouble. Most of the subject seems quite obvious to the lecturer. So the lecturer tends to neglect the third fundamental rule of teaching, to explain the obvious. Thus the lecturer on error correcting codes did not explain the strategy for correcting errors.

Many of the mistakes I note above are committed by newly graduated mathematicians (but they are by no means the only culprits). Part of the problem lies in the fact that in a seminar one must show that one is very clever. But I feel the success of a seminar lies in how easy it makes the subject appear, and not how hard, and a seminar should not be a way of selling oneself.

Of course, it is also possible for a lecturer to signal what the listeners are expected to know in the abstract to the seminar, e.g. by writing "Knowledge of elementary coding theory will be assumed, for instance Part 1 of 'Error-Correcting Codes and Finite Fields' by O. Pretzel", so that the lecturer does not need to define the fundamental definitions.

7.4 Some extra remarks

If you ask me as the author of a book on how to teach mathematics, what you should do when you have only sufficient time to work on some new mathematics or else to read a book on teaching, then my reply is unequivocal. Learn the new mathematics. The more you know about a subject, the more you have to teach, and even if you have poor methods of teaching, that must be better than teaching a restricted number of ideas well. But hopefully you will have time to study teaching as well.

What then about various methods of teaching? Should one go for teaching individuals such as in reading a book, or as in the Keller System, or in small groups, or project work, or problem based learning, or programmed learning, or teaching in large groups for instance (see §10.6 for a brief discussion of these various methods of teaching)?

Should one favour continuous assessment or an annual examination? There are many controversies about these things. I favour the following approach. Don't be dogmatic. These various methods should be regarded as a number of tools in a tool box. Each has its advantages at the right time. Select the right tool as and when it is needed. None of these systems is superior all the time, just as it is impossible to claim a screwdriver is better than a hammer all the time. If you have the time and the energy, find out about these systems, and try them out. Often the main value of a change to one of these methods of teaching is that it represents a change, which is an advantage just for the variety which it brings.

PART III DEPARTMENTAL MATTERS

Chapter 8	Organisation and Examinations
Chapter 9	Planning
Chapter 10	Methods of Teaching and Equipment

8. Organisation and Examinations

Preview

This chapter is aimed at emphasising the importance of organisation in education. By this I mean such prosaic things as the arrangements of the terms, the examination periods, the re-sit periods and so on.

Unfortunately, organisation is often decided by the central administration, and is decided without sufficient thought being given to desirable education aims. I once met a group of principals about to set up new higher education establishments and I tried to persuade them of the importance of these issues. They were very polite, but I realised that I was wasting my time. They had other, more important things on their minds, probably politics and finance. Yet the main object of the universities they were setting up was teaching.

8.1 Organisation

What is desirable in organisation is often obvious, but is usually difficult to arrange, mainly because of the difficulty of planning a number of measures which may conflict, or because there are not enough staff, or lecture rooms. For instance, administrations are very keen on a modular system, so as to maximise the students' flexibility of choice. Mathematics courses, which often depend on a sound knowledge of previous courses, are thereby forced to include extra bridging material, which reduces the material that can be covered. Examinations which are spaced so that there are intervals between them are easier to pass, and involve less stress, and so on. I think it is time for departments to take a much greater part in deciding how these administrative matters are organised.

There are many ways of organising degrees and degree structures. In this book I do not wish to get involved in discussing these, but content myself with saying that such discussions are well worth having, and also refer the reader to LMS [1992] for a discussion of various systems of organisation in the report "The future of Honours Degree Courses in Mathematics and Statistics."

The way the academic year is organised will have a very important effect on study. How will the courses be studied? Is one to study only one course intensively for a period, or is one to study two or more courses in the same period? Research is needed into this point. Departments should really consider these possibilities consciously, rather than simply accepting the status quo. At different universities one, two, three or even four courses are studied in the same period.

Studying one subject intensively has the advantage that the conflict for attention that arises when two or more courses are studied does not arise. Also one can achieve a great deal in a relatively short period. Against that, one does not have a chance to reflect, to allow the material to be absorbed.

With two or more courses one has an opportunity to vary one's study, changing to another topic when the mind is exhausted with the first. Also each course is then studied more slowly in time, thus allowing a maturing process to set in. All of these are of value.

However, then it is difficult not to place sudden periods of strain on the student, when several courses require extra effort. So whenever several courses are involved simultaneously, ways should be found to co-ordinate the courses, so that difficult and time-consuming periods in different courses do not occur at the same time. I have never seen an attempt at achieving this made, it being one of those jobs that is never assigned to anybody.

The placement of the examination is important. Should all examinations occur at the end of the year, or after each course is concluded? What provision for re-sits should there be?

There are some universities which hold examinations all together at the end of the year. This imposes a very demanding discipline on the students, requiring them to work consistently through the whole year, although there are always some who coast the whole year and manage nevertheless to pass with a final desperate cram at the very end.

Other universities impose an examination every term, thus forcing the students to work steadily throughout the year. This also means that the examinations can be more thorough, since there are more examinations per subject.

Of course having examinations after each course, rather than at the end of the year, does make it easier to pass.

The Swedish policy concerning re-sits is very good. It allows each student three attempts at each examination, and the second attempt is often quite soon after the first, giving the student a significant extra incentive to study after the first failure, and ensuring the student will revise while much of what has been studied is still fresh.

A well-designed time-table is of value in teaching. Such a time-table should be regular, not varying from one week to the next. It should take account of good teaching principles, namely there should be a time to introduce the new topic, to go through it in more detail, to do problems, to have an opportunity to work with fellow students, and to have a chance to ask questions of the teacher.

The work should be evenly spread over the year, and if there are simultaneous courses, hard periods in the one course should be balanced with easy periods in the other course.

Time should be arranged to prepare for examinations and re-sits. It is of course difficult to make these arrangements; the more centralised the organising of the time-tables the harder it is to check on these things. I am therefore keen on having the time-tables organised by the individual departments, rather than by the university.

There must be some procedure for measuring the overall load on the student. This is particularly important where there are two or more courses proceeding simultaneously.

8.2 Examinations

Are examinations mainly to assess students or to help them learn? Personally I feel that their main value lies in helping learning. Being forced to revise a topic again means one gets a very good understanding; indeed, it is often only on the final revision that students really begin to understand the subject. Deep and sophisticated ideas take time to sink in, and require repetition and contemplation to be absorbed.

However, as a bonus, examinations are also used to assess students, and to finally grant them a degree. It is therefore essential that they be just. By this I mean they should cover much of the syllabus, and they should constitute a reasonable test of ability, and that marking should be fair, with all students who produce the same answer getting the same number of marks. There should also be an appeal system.

In the case of written tests, it is easy to arrange an appeal system. I have my method. First I genuinely try to see when requested whether I can increase the mark, by looking through the examination with the student. It therefore helps to have an explicit marking scheme to which I can turn, in order to show the student that marks have not been awarded arbitrarily, but have been decided carefully beforehand. If I find that the student is entitled to extra marks then I willingly give them. Otherwise I try to explain why none should be given. Then if the student is still not satisfied I suggest we give his paper to somebody else to check. If the other marker manages to get a pass for the student then instead of arguing I simply agree to the result and congratulate the student. In the case of oral examinations, it is much harder to make the examination just.

It is a truism that the examination steers the student's activity; indeed it is often the main mechanism. A time-honoured and very effective way of revising is to go over a number of old examination papers. When I was at Lund I faced five extremely determined and angry women students. The reason why was that I was intending to provide only three old examination papers for them to revise and they demanded ten.

Since students spend so much time and effort studying old examinations it is important that the examinations over a period cover all the important aspects of the courses.

Providing one or two tests during the course can be helpful in getting the student to revise. It gives some high points in the course, a chance to see whether the methods of studying so far are successful or whether they need to be changed radically.

On the other hand, tests mean that the student is being directed, that this is more like being trained rather than educated, which runs counter to my comments in §8.5 below on spoon-feeding. However all of these comments on teaching have to be balanced one against the other. Then too the incredible amount of time required to set and mark tests needs to be taken into account. The desirability of any measure in teaching has, as in ordinary life, to be constantly judged by its expense.

Tests need to be carefully controlled so that they do not lead students to feverishly study subject A, and having written the test, stop studying A so that they can study for the test in subject B, which they then neglect while studying for the next test in subject A. This yo-yo effect must be avoided, and it requires careful co-ordination and co-operation between lecturers and departments, which is not easy to achieve, but is nevertheless of importance.

I have two main experiences of oral examinations.

My main experience was in Lund. The students first had to pass a written examination, which consisted mainly of computational questions. The successful students were then examined orally on the more theoretical aspects. The examination was carried out in a number of small rooms, each with a blackboard. Each student was given a question, and asked to write a solution on the blackboard. When ready the student would then open the door. This was a signal to the examiner, who was waiting in the corridor. Normally, the examiner dealt with two or three students simultaneously, and so would be kept busy most of the time.

The examiner would check the solution written on the blackboard, and ask further questions. Quite often an apparently perfect solution had not really been understood, and the questions revealed this. The student learned a great deal from this questioning.

I remember in particular one woman student turning to me in exasperation and saying "I don't know why you keep on asking me these peculiar questions." The questions were actually rather standard. "Give an example where if the conditions in the theorem are not satisfied the conclusions do not hold", for instance. I failed her, and she had a re-sit a couple of weeks later. She had learnt from her experience, and passed very well. Incidentally, when re-examining, the examiner was changed, it being too much of an ordeal to be examined by the same person twice.

The oral examinations not only develop the ability to explain oneself orally, they also lead to asking the student a number of essential questions, and also to the ability to think on the fly, and, to discussing the subject in a wider perspective. I have always noticed that English graduates seem much less able to express their thoughts verbally than their counterparts from the Continent, where oral exams are regarded as important. Oral examinations are however very time-consuming, and again their desirability must be carefully balanced against their cost.

My other experience of oral examinations took place at Mälardalen University. In this case I felt the oral examination was not very satisfactory. This was first of all because the actual course being studied involved much discussion and individual questioning so that most of the advantages of the oral examination had already been acquired. The purpose of the oral examination was therefore simply an examination, and in this respect it was a more subjective and haphazard a process than a corresponding written examination.

There are students who have learnt a fair amount and yet in the written examination get virtually no marks. As I point out to them, it may not require very much more effort to get a really significant increase in their marks. I say that the graph of mark obtained against study time looks like Fig. 8.1.

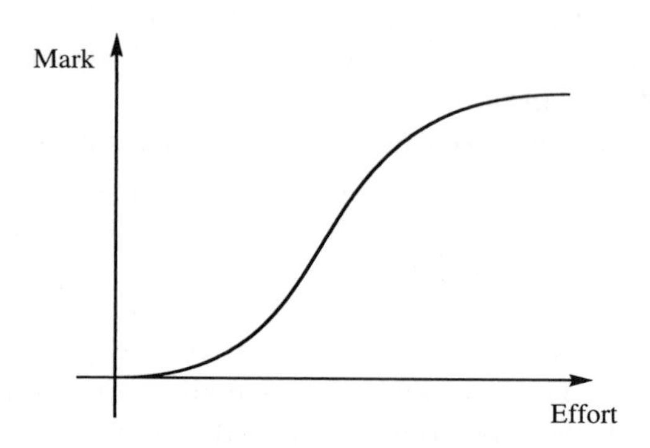

Fig. 8.1 Mark versus study time

That is to say, the number of marks obtained may be very few despite a lot of work, and then the number of marks begins to increase quite rapidly with only a modest increase of extra work.

There may be a point in trying to find some other sort of examination to encourage the students who score few or no marks on the regular examination, despite their having learnt something. For instance, an oral examination which allows a certain amount of prompting, will enable students who have a more passive knowledge of the course to illustrate their knowledge. This may help their motivation. It is very dispiriting if one has worked hard and nevertheless obtained very few marks. It is also not impossible to think of a written examination which can give extra prompting. The student would have to decide which examination to answer, the one with prompts or the standard examination.

For example:

Standard question: Calculate $\int x\sqrt{(1-x^2)}dx$.

Prompted question: Calculate $\int x\sqrt{(1-x^2)}dx$ using the substitution $u = 1-x^2$.

With a suitable interactive computer program, it may be possible for each student to vary the difficulty of the examination at the time of the examination.

Exponents of continuous assessment point out that many students postpone their studies until the final examination, and then try to learn the material in a hurry, giving rise to poor knowledge and understanding. Continuous assessment, on the other hand, keeps the students working steadily.

Like almost everything in education, these remarks have a point. On the other hand, it is easier generally to pass examinations based on continuous assessment, and there is the danger, if there is no final examination, that the student may forget the earlier parts studied, secure in the knowledge that those parts of the course will not be examined again.

Also many students find it easier to start studying the course seriously only after their lecturer has gone through all the material. In other words, first they get a rough overall picture of the course, and then they feel they are in a position to study the course. For some students this is a very much more effective method of studying than learning each point step by step as they go along. Others of course, find it best to learn each step in order.

One of the advantages of a final test is that the student has an incentive to go through all the material, and relate the various parts to one another, and in that way get an overall view of the course. So I am in favour of a final test, even for courses where there is continuous assessment.

It is advisable if some time can be allowed for diagnostic and revision tests. They must actually be time-tabled, otherwise in practice it is difficult to fit them in. The diagnostic test could sensibly begin each course. It is useful in indicating which topics the student has problems with, but perhaps its greatest value is in getting the student to remember the important topics from previous courses which are needed in the new course. For this reason I am in favour of giving the student a sample diagnostic test, and making the proper diagnostic test a virtual copy. Such a test could consist essentially of the important definitions and statements of theorems from the previous course which are now needed in the present course.

It is also a good idea to arrange at least one test during the course. This will help cement some of the ideas in the mind, and also serve as a warning if the student is not studying properly. However, it is worth while keeping the remarks on directed teaching in §8.5 in mind. Some direction especially in the beginning is helpful, too much is not, because the students will never learn to control their own studies.

Project work requires the ability to organise and co-ordinate written material. It is a different skill to writing examinations, or performing well in an oral examination, and some practice is invaluable. It requires the student write out the argument carefully, and this is then carefully criticised by the member of staff, leading to a second draft, which is again criticised, and so on. The difference between the first and the last draft is usually very impressive.

However, all of these take time and cost money, and it is not clear whether they can always be provided at all institutions.

8.3 What exactly is your examination supposed to test?

It is worth while spending some time thinking of what precisely your examination is testing. This could be perhaps divided into the following:

- Knowledge of definitions, theorems and algorithms.
- Standard applications.
- One-step problems
- Two or more steps problems
- Problems which illustrate ingenuity
- Understanding of the subject matter as a whole.

The attainment levels in §2.5 and the ideas in §13.9 could be used to help analyse the examination. Useful ideas on examination can be found in Schoenfeld [1990] and the references given there under the heading "Evaluation", in Burn [1997], and also in Griffiths [1974].

8.4 A much neglected teaching aid

Immediately after the examination is over formally, each student should be provided with a solution of all the questions on the examination. This should preferably be very carefully written, with clear and proper explanations, and references where needed to the text-book. At no other time is the student so attentive to mathematics. (This is less true if this is the student's final year of examination.)

This simple method teaches the students a great deal. However, it must be done properly. It is not sufficient to write a quick and abbreviated set of solutions to the problems. The work should be as carefully set and thought out as if it was to be published. Staff must be given extra time to do the job properly, and it should be checked by another member of staff. (At some universities checking by another member of staff is sensibly done before the examination.) These solutions should also be made available to students in following years, so that they can study past examination papers.

8.5 Directed teaching (also called spoon-feeding)

All teaching is by definition directing the student. There are various degrees of direction, and the higher the degree of direction and the more compulsion the closer we are to teaching as practised at school. The following levels give an idea of what I mean. Each level is taken to include all the measures in the previous levels; thus for instance degree 4 includes all the items mentioned in degrees 1 to 4 inclusive.

Degree 1: The subject is introduced with some overview and various text-books or other sources indicated.

Degree 2: Detailed lectures and instruction are given, so that all the relevant material is covered.

Degree 3: Tests are organised so the student is given an aim to work towards.

Degree 4: The students are checked to see whether they are working steadily and are understanding the material, and if not, corrective measures are taken. The discussions of the material with other students and the teacher is organised by the teacher. The revision in the form of tests or review questions is organised and controlled by the teacher.

Degree 5: All the above but compulsion is now introduced. The student must be present at class or lectures or else faces a penalty, such as not being allowed to enter the examination. Not completing all assignments will result in failure, and so on. With school children they used to face some punishment, like staying on after school was over, to do some routine task, or write lines, or else they would (long ago), be caned by the headmaster.

I have no objection to what the student does under all these degrees of supervision (with the exception of punishment). All of these procedures are excellent for ensuring that the student will learn. What happens is the higher the degree of direction, the closer you come to school teaching, with a mixture of instruction, exercises directed by the teacher, with student discussion being organised by the teacher, and then the students being required to present a little talk on the subject, then the teacher summarising and giving a further exercise and so on. Such a form of teaching is described for instance in Ramsden [1992] page 179, where it is advocated as a method for teaching at university level.

The directed teaching systems do certainly entail all the ideas I have been espousing: active work by the student, revising, problem solving, discussing with fellow students and teacher. These are very good, and my objection to them at university level is that they are orchestrated by the lecturer and not the student, whereas it is the student who should be arranging these things. The student must at some stage of his development learn the techniques of learning and be able to organise his own studies. So if you do use a more or less directed system of teaching at university level, point out explicitly to the student precisely what you are doing, so that he or she can learn and adapt these methods to organising his or her own learning in the future.

A second objection to directed teaching is that the process becomes incredibly slow. If the students can themselves apply the techniques of good learning they can in their own time master the material. Otherwise the course goes along at a very slow rate. Different things stop different students, and the net result is if one has to wait for all students to master each part of the course, one will have to wait a long time, and only a fraction of the desired material will be covered.

Now it may be objected that a large number of the suggestions I have made previously and will make subsequently are simply this sort of directed teaching, also called less kindly "spoon-feeding." Thus, the tests and examinations I have advocated in §8.2 above do organise the student and his revision. Am I then advocating spoon-feeding?

No, I am not. Some directed teaching is helpful, especially at the beginning of the student's university career. It is a matter of balance.

Eventually the aim is to wean the student of this extra help, because the student must be taught to take the responsibility for organising the work. There are those students who can not organise themselves, those who really would not have been admitted to a university some twenty years ago; but yet, it is now a fact that they are at the university and it is better to get them to learn something by using directed teaching than to leave them to sink.

Consequently, as many of the students now coming into the university are not very good at organising themselves, some organising must be done on their behalf. But gradually the degree of direction should be decreased, and the students taught to cope with the problems of studying and organising their study themselves.

That is, as I have claimed before, the aim: to produce graduates who can themselves direct their own studies and learn new ideas and material, perhaps ideas which were not even known when they were at university!

The degree of directing the student should never be so high that the student can never make a mistake. It is an excellent lesson to neglect one's studies, to fail to do the things one knows one should, and then come a cropper. Students should be allowed the luxury of at least one serious blunder. That way they learn.

A popular method which usually results in better student performance is a very detailed and clear description of exactly what is required for the course. Thus one explicitly gives the student a list of what should be learnt, for instance: "You must know how to calculate the scalar product of two vectors, and know how to find the distance from a point to a line. You must be able to calculate the cross product of two vectors. You must know how to prove Theorem 1, but it suffices to know the statement and to be able to use Theorem 2, etc."

This is a helpful method. My objection to using it routinely is that one of the hardest problems in life is to know what exactly one has to do. By giving the students precise instructions you never give them practice in finding their own methods of working out what they have to do. This skill may very well be more valuable than actually learning the mathematics. So although I do not object to this method being used in some courses, it should not be used in all courses.

8.6 External examiners

In England there is an elaborate system of external examiners. A mathematician of standing from another university is chosen as external examiner. The job is to firstly judge the standard of each examination paper, and then check whether the marking has been carried out justly. This involves a great deal of detailed, individual work. Typically an external examiner will be given the examination papers in advance, and will comment on the questions. The external examiner will also examine examination scripts which lie on the border, and will also decide on the final rating.

The system has a number of advantages, ensuring that the universities will broadly speaking be comparable. No lecturer wants to be shown up in front of the external examiner, and consequently takes considerable pains when preparing an examination, so much so that the external examiner seldom needs to make any changes. The whole procedure, however, involves an enormous extra administrative load.

Examinations in the USA can be gloriously informal. I recall one lecturer who told the students assembled for the examination, that those prepared to settle for a B could leave; the others would be subject to an examination, which would give them a chance to gain a better grade, but of course they might then fail. Half the class left. Whereupon, without further ado, the rest were awarded an A. This is of course not typical of the U.S.A. but shows what can happen if there is no external check. In my own university in Sweden there is no external examiner, and this too can lead to abuse. There is not even the precaution of having another member of staff check the examination. This I think must be the minimum required.

8.7 Summary chapters 1—8

Chapters 1—8 provide a framework for the discussion of teaching at universities, and it is apparent that there are a host of contradictory desirable measures that need to be considered. Fundamental to all discussions is that that everything costs time and money, and one will need to set priorities. Lecturers are very hard pushed to do their many tasks, and can consequently only carry out a few improvements. There is also a limit to the improvements that changes in teaching methods can produce.

I have come out strongly in favour of each university aiming to produce a number of outstanding graduates. There is a very real danger that this aim will be ignored since the universities are being swamped by quite ordinary students, who find mathematics very difficult, and we must make sure that they prosper. This can only be achieved if they have suitable courses which take account of their abilities. Meanwhile the more technically talented students must have more demanding courses; it is not possible to teach all the students simultaneously, in view of the huge variety of skills and abilities, and we will need to have a large number of different courses, and of course, more funding per student rather than less.

It is important nowadays that the less capable students are catered for, and that the aim is that everybody should leave the university with greater skills, that is, one comes to the university to learn, not to fail. As to whether our courses should be directed or not will depend on the students involved, but the aim will always be to get them to become more independent and more capable of directing their own studies. There is thus a need to directly teach the student techniques of study.

I have pointed out that the provision of good teaching can be facilitated if we all work together, that it can not be left to the individual teacher. Improvement in teaching is going to mean attendance to a large number of bread and butter items, and will not be achieved by a radical new approach. In particular, it is worthwhile stressing the advantages of teacher-centred learning.

Now there is no doubt that governments have had and have always had an important say in the provision of education. It is important that they get the education decisions right, since mistakes in education take decades to correct. Countries can easily and rapidly decline if they get these decisions wrong. Many governments see universities as mainly a chance to correct inequalities due to class and race, and also a way of keeping unemployment figures down. This is the same approach which has been used to control schools. It has led among other things to the more able students marking time at school, instead of learning as much mathematics as they can. Schools and universities have among their main aims, the transfer of ideas and learning, and governments should not forget this.

Universities need to form research units to decide what are the most likely and useful subjects and mathematical techniques to study at university. I see a lack of interest in governments to consult the lecturers, the teachers, the schools and the universities. Part of the problem is that university lecturers do not spend much time considering education in general. We need, all of us, to sit down and think. I am convinced that there is nothing more wonderful than education and that it will be our salvation in the future.

9. Planning

Preview

How shall all the comments and ideas mentioned in the preceding chapters be put into practice? Although there is no sharp division, there are some things which can be arranged and done best by individual lecturers and these are discussed mainly in Chapters 11—13. This chapter and Chapter 10 are mainly concerned with matters which benefit from the co-operation of the whole department. It is sensible if each department can have somebody whose function is to see the structure as a whole and co-ordinate.

9.1 Fundamentals of departmental organisation

The department will need to realise that not all desirable objectives can be achieved, and so compromises will be required.

One should certainly consider the infrastructure. Of course this really is not under departmental control, and for this reason most departments remain quiet about obvious defects. Consequently nothing ever improves.

Of course the halls should be properly designed with teaching in mind. If you have small groups, you cannot succeed with a room designed for 60 students, for instance. If you have 60 students, can all of them see and can all of them hear? Is the lighting adequate? Are the blackboards (white-boards) sufficient? One of the most effective improvements to teaching is the provision of more blackboard space, with good lighting. Are the seats reasonably comfortable? Is the room kept at a comfortable temperature? Is the ventilation adequate? All these things are obvious, yet often they are neglected and students' learning is thereby adversely affected.

It is apparent that one's choice of system of teaching is dictated by the lecture theatres. If one has a theatre which can seat 200 students, one can consider lectures for a large group, and then divide the students into smaller groups for tutorials. If however one does not have such a hall, but only has a hall for 100 students, then one would have to organise two different lectures of 100 each. This means that important decisions about the teaching are dictated to us by the original group in charge of planning the building. This suggests to me that there should be considerable consideration and wide consultation when the buildings are in the process of being planned. Architects designing university buildings should try also to leave some degree of flexibility in the rooms they design. For instance, a little extra thought and trouble should lead to the possibility that the lecture theatres could be divided in two later on, without much extra expense.

In my lectures I see the students at the back struggling to see over the others. Yes, it is expensive to arrange to raise the last two rows of seats, but taken over a period of 15 or so years, the cost is surely negligible? It is worth while putting in sound absorbing material to improve the acoustics. It is worth while putting in new lights.

These things are not attended to mainly because they are not the responsibility of any one person. Lecturers and students may complain a little, but seldom take the matter further. And the administration who should take care of these things never do. Often they are not even aware of them. Their attitude may very well be different if these matters are brought to their attention. For instance, two of my colleagues, pointed out firmly the need for large white-board space before the final specifications of new buildings were decided. With not much difficulty they succeeded in persuading the administration to install movable white-boards. This one step has improved the mathematics teaching enormously.

It is common at the end of each course to have a questionnaire. Add some extra questions to this questionnaire asking about these practical matters. Make sure the administration gets a copy of the replies.

An audit of these facilities can easily be arranged, say once a year, and one can easily find out the weaknesses and strengths of the lecture theatres. The next thing is to do something about them. But the first and major step, is for there to be somebody responsible for these facilities, who takes these matters seriously. More generally, I am saying there must be some way in which the experiences of the students can be fed back to the decision makers.

A strong staff/student committee is very desirable. But it is also the attitude of the staff that is important. They must take the students' views seriously. I take it for granted that each class should have a student representative, who can put the students' points of view strongly to the lecturer, and if this does not succeed, go on to discuss the matter with the director of undergraduate studies, or the head of department. Each lecturer should as a matter of routine arrange to meet the class representative regularly, say once a week. If all goes well, such meetings take only a few minutes. Otherwise the lecturer learns of problems early on. This is something that of course can be arranged by each individual lecturer, but if it is an accepted part of the department's approach, it works better.

9.2 Mentors for students and staff

Mentors for students involve the department in considerable extra expense, but is certainly worth while. Each student should ideally feel that there is at least one person in the department on his or her side. If the student can meet a mentor regularly say for 15 minutes each week during a whole term, the mentor can be turned to with confidence in subsequent terms. Such a scheme functioned quite well at Imperial College. It was the custom for the student to meet the mentor on the first day of arrival immediately after going through the formalities of registering. The student would then be invited to lunch, and further arrangements would be decided then.

It is even better if the mentor and the student can work on something together, preferably something outside the ordinary tasks that the student is supposed to tackle, so that they have a chance of both contributing. For this purpose it is of value if they discuss something that is unknown to both of them. The lecturer could well begin by suggesting an historical article, or a mathematical article well within the intellectual range of the student, who could then be encouraged to research the matter, finding new material, and trying to explain these new ideas. The Mathematical Intelligencer, the Mathematical Gazette, and Scientific American are excellent sources. The idea is to choose an article which has a large number of references, so the student can research them going backwards in time and then write a project on the material. This would provide further help in teaching the student verbal and written expression.

It seems also sensible to have a student mentor as well, somebody who can see that the new student's social life is helped along. Particularly in the very beginning, students can be lonely.

A small item, which is easily organised and helps both to reduce the secretarial load, and also helps an uncertain, beginning student, is a student's manual. By this I mean a small booklet with many of the practical details, how to register, take examinations, dates, who to speak to for advice, how to appeal about examination results, how to complain about lecturers, etc. This is useful to prevent the student wasting time. It is not necessary even to produce a booklet, the information can be available on a bulletin board, or else on the home-pages.

While we are discussing it, a lecturer's manual is also a great help. Such a booklet would include all the minor organisational details needed for the job. The lecturer needs to know how the term is organised, when the examinations take place, where to get his stationary supplies, which examinations he must prepare and mark, etc. Most new lecturers waste time, theirs and the secretary's, finding out about these obvious features. And yet if one knows these details, it is easier to plan one's time efficiently, and enable oneself to be on time with everything.

I also think that each beginning lecturer should have a mentor, somebody to turn to ask for help and guidance. It is after all a considerable strain to suddenly take on teaching without any advice or guidance. Help and reassurance is required. Yes, after a while the new lecturer will be coping well, but in the mean time his students suffer and he has difficulties. Even an experienced lecturer coming to a new department should have a mentor. But a beginner definitely needs one.

The best mentor is one taking a parallel course, or somebody who has taken the course before. The mentor does not need to spend much time on his task, needing only to tell the new lecturer when to get ready for the next stage of preparation, and answering any questions that arise with the new person's work.

9.3 Department leadership

The head of department has a number of important functions, which involve a large number of administrative, political and financial problems, as well as academic and personnel problems. The job is demanding and difficult.

Nevertheless I feel the head of department ought to take a closer interest in the staff than is normally the case. It is an important part of the head of department's functions to encourage the staff. The most successful head of department I have met told me his main aim was to try to obtain promotion for his staff. He worked very hard on that. In consequence he had an extremely happy, hard working, optimistic staff, and inevitably, the quality of teaching, co-operation, research was extremely high. In other words, choose the best possible staff you can, and then encourage and support them, and then you need not do anything else. The staff themselves will do the work, improve both their teaching and their research, think of new ideas. They will work a lot harder than if they are ignored, or threatened.

Most heads of department just leave one alone. And thank goodness for that. What is the point of having staff you have to organise all the time? But leaving staff to get on with things on their own is not sufficient. I have seen departments filled with unhappy, discouraged staff, many of whom have given up trying. They have never received the slightest encouragement in twenty years of lecturing, not even a gesture of appreciation, just complaints. No wonder bitterness sets in. Their early promise and prospects have faded. The problem is also that there is nowhere to advance in the academic world. One seems to remain always in the same position. Indeed, it's worse than that. People are described as "dead wood". Far from being appreciated, they are regarded as a disadvantage to the department. In the long run that is dispiriting. In the long run, as a consequence, they become dead wood!

I am much in favour of democratic decisions. I recall a head of department who in fact based the final decisions on a democratic vote of the department for anything of importance. First the matter was investigated by a small committee, which presented its findings, which were subsequently discussed by the department. Finally there was a vote.

The head of department should be concerned for each and every member of the department. He or she should take the time to speak to each of them to find what their aims and ambitions are. An occasional word of thanks for work done, congratulations on a paper just published, or a remark acknowledging good teaching, all help morale.

The mathematics departments of this world are full of very clever people, and surprisingly, many of them have no social skills. If one is head of department, every so often one needs to do something a little difficult, maybe even unpleasant. Such heads of departments should study how easily and skilfully people in very much less intellectually demanding jobs manage to cope. It is a matter of putting oneself in somebody else's position, empathy in fact, a word few heads of department know. Very clever and ambitious, perhaps even ruthless people, can find this difficult. Charm, kindness, a little humour, and respect for others, can make it very much easier to handle personal difficulties. Maybe if you do not have these qualities, you should think twice before becoming a head of department.

A true test of a head of department comes with a retiring member of staff. When one is retiring one is filled with mixed emotions. It is time to go, and yet when you consider your working life you are not too sure whether it has been a success. Your colleagues, who do not need you now, begin to show their true feelings.

Usually when you retire you have reached the age when you are not doing research. The head of department who is ambitious for the department does not really think much of you: the staff he or she likes are those who are producing more and more research, and the head of department finds it difficult to conceal the pleasure felt that at long last you are leaving. He or she is busily and happily planning who your replacement will be. And yet a member of staff who has worked hard for the department needs a little respect. And that respect is something which gives many benefits to the rest of the department. A department in which the staff retire in bitterness and anger is not a happy department, and not a department that can do good research and teaching, and that bitterness will not encourage loyalty.

Aside from being somebody who keeps the staff in good spirits, the head of department should spend a considerable amount of time thinking of the future, the future of the research in the department, the future of teaching, the further developments which can be expected. The head of department should try to be several steps ahead.

I find it easy to imagine a very good head of department who is a poor administrator but not a head of department who is good solely because of good administration. The head of department should also communicate ideas up to the rector or vice-chancellor, who should in turn tell the education department and the minister of education of the thinking in the university.

9.4 Timetables

It is important to have some method of checking the total load of the students from one week to the next. Account should be taken of difficult periods in concurrent courses, so that two or more subjects do not simultaneously become very demanding. Particularly demanding are tests, which can lead to students ignoring the other courses for a short period.

To study the total load is relatively easy in the Swedish system, where each student is given a detailed planning of the course in advance. The next step required is to look at the detailed plans as a whole, and modify the plans so as to avoid bottle-necks. It is this step which in practice does not seem to be carried out. Each lecturer sticks to his own plan, and does not bother about anybody else's requirements. Yet a brief meeting of the lecturers who are involved could result in shifting a test back or forward a week to avoid an excessively tough period. Even in England where there is seldom a formal requirement for such careful and detailed planning, a brief meeting each week of the two or three lecturers involved could easily avert problems. Again, this is something which is not done, not because it is intrinsically difficult or time-consuming, but because it has not been done in the past.

This sort of organisation is harder to accomplish the larger the university and the larger the choice of options. But it possible to think out something: e.g. courses are divided into say two mutually exclusive groups and those in Group 1 can only have more extended tests and assignments during odd weeks, and those in Group 2 can have them only during even weeks.

It is also worthwhile asking the students whether there is somebody willing to make a record of their work during the weeks . Some students quite like this idea of keeping a diary, and will do it conscientiously without feeling it is a chore. The results from one year can then act as a rough guide in determining what is to be done the following year.

The department should check that timetables are well thought out. It is usually best if there is a degree of regularity built in to the system, and if examinations and re-sits are also considered.

It is sensible to arrange to present new material to students, then give them some time to go through the new material and do some problems, then have a chance to ask questions in a tutorial. Avoid such obvious mistakes as a having Monday's lecture 5-6pm and Tuesday's 8-9am. The student must have time for reflection.

Different lecturers have different methods and stress different approaches. It is important that these different approaches are experienced by students. For that reason, it is a good idea to try to ensure that each group of students will be exposed to a large variety of lecturers. It is also worthwhile changing the courses taught by lecturers at reasonable intervals, because it is difficult to maintain enthusiasm for a particular course if one has taught it more than four years in a row.

9.5 Study skills

The teaching of study skills, as already mentioned in Chapter 6, should be built into the program. The whole department will have to decide how this is to be accomplished and who is going to carry it out.

The whole department will have to decide on the larger variety of courses that will be necessary now that a wider range of students are coming into the university.

9.6 Deciding which are to be methods courses

The beginning courses at university I feel must be methods courses. Instead of proofs, one needs demonstrations. Take for instance, the derivative of a function of a function. It is possible to write out a simple "proof" which is convincing but may be wrong because one may be dividing by zero in a particular case. Because of that one has to be a little more careful. The result is that the proof is more difficult and looks less plausible. There is nothing wrong in adding an extra condition parenthetically if you do not want to state something false, but why not stick to the easier demonstration. Again, the various levels of proof described in §2.5 are relevant. One should gradually take the student through the various levels of proof described there.

For students who are going to specialise in mathematics there must be a number of carefully developed courses using an axiomatic, definition and theorem approach. However, not all courses need to have the same degree of rigour. The department can sensibly decide which courses are going to be more or less precise. In §10.1 we take up the important question of how the axiomatic method of argument can best be introduced to new students.

9.7 Summary

This chapter was mainly concerned with stressing the need for a departmental approach to certain matters, most of which are plain common sense, but tend to be neglected because they are nobody's responsibility. The next chapter is also one in which many of the decisions to be taken need to be taken on a departmental level.

10. Methods of Teaching and Equipment

Preview

This chapter continues to discuss matters that must be decided by the department as a whole. It is important for instance to decide where and how and when the definition/axiom/theorem approach should be introduced, and which courses should be studied in that way. It is also sensible to discuss which methods of teaching will be used so that the student can be exposed to a large variety of methods. It is essential to decide the role the computer will play in the course as a whole.

The provision of blackboards and or/overhead projectors and consequent effects on teaching should not be neglected.

10.1 How to introduce the axiomatic method

First, let me point out to pure mathematicians that it is not necessary that all courses be carefully and logically correct, with development from axioms. In fact, since this process takes a long time, one can learn a lot more material by taking a more cavalier approach. In this matter pure mathematicians have much to learn from applied mathematicians and physicists, who tend to take a pragmatic view. It is important to have several courses which properly develop the subject, but if we want students to know a fair amount, then we must learn to give more methods type courses. Pure mathematicians will have to restrain their instincts. So it is important that the department decide which courses are to be given in a strict fashion and which are to be given as methods courses.

Nowadays it is not uncommon for students to first meet the axiomatic method in a course of analysis or a course of linear algebra. They are completely thrown, and it is not surprising. They need a chance to develop a feeling for axiom, definition and theorem proofs. This I believe could be better achieved by some of the following approaches:

- One could develop the complex numbers assuming the real numbers.
- One can develop the rational numbers starting with the natural numbers defined through Peano's axioms.
- The beginnings of the theory of projective planes could be taught.

By proceeding in a gradual fashion, one hopes to slowly introduce the student to the axiomatic method, with definitions and careful proofs. One must point out the value and interest in such an approach. Now whether these various approaches should appear in a single course or in a number of different courses which then can be combined so the students' approach to abstraction is facilitated, is one of those matters which require departmental organisation. One must also insist on the value of trying to combine both intuition and rigour.

Students need to be convinced of the value of rigour. This can be done by arguing intuitively and obtaining wrong answers. (Limits provide good examples of this, as well as series.)

A helpful analogy is that of flying an aeroplane. One checks the instruments carefully and follows the directions on the instruments. However, it is reassuring to look out of the window and see the ground, estimate one's height and speed and so on. Nevertheless, the instruments are likely to be more reliable. When flying blind they are indispensable. When you cannot trust your vision, when you are uncertain, you know you can rely on the instruments.

Similarly, using axioms, definitions, and theorems rigorously is like flying blind: you know it is a safe and reliable process, but if you had some intuitive feeling you would feel much more comfortable. In the study of mathematics we need both the security of the axiomatic checking and the intuitive feeling.

10.2 Techniques of study

We have already mentioned techniques of study in Chapter 6. These various techniques need to be introduced, repeated, and re-emphasised several times in several different courses if they are to be of use. The department will need to decide which courses will teach and emphasise these methods. It is ineffective to introduce them once and for all and expect the student to manage after that.

10.3 The role of books

Many courses are today taught using a set text. The advantages are quite clear: the detailed information comes in the book, thus freeing the lecturer to present the material in a complementary fashion. It does not need to be presented in a logical order, after all the book does that, it does not even need even to be accurate if a rough argument rather than a precise one can convey a better intuitive understanding. I call this the "text-based" method.

The other method, which I call "the lecture-based method", does not use a text-book, but the lecturer will give detailed and clear notes, and will often refer to a number of books which students are urged to consult. The advantage of such a course is that the lecturer can tailor the course to the students' exact requirements, and produce a shorter and more concise course. The disadvantage is that both the student and the lecturer have to spend time just recording the information. That has its advantages too: it is less likely that a too extensive course will be given, and the very effort of copying down the material gives a preview of the material. It means too that the course can be highly flexible. If one has chosen the wrong book with the text-based method, it may prove to be either too hard or too easy, and there is not much one can do in these cases.

There have been some remarkable (but mainly superficial) changes over the last thirty years in text-books. Firstly their lay-out has improved immeasurably, with beautiful diagrams, pictures, headings, the use of colour. For instance, if one compares Hoffman [1965] "Linear Algebra" with Norman [1995] "Introduction to Linear Algebra," there is a huge difference.

It is clear that the older book is much more ambitious in the topics covered. The layout, although clear, lacks colour, and is not as generously spaced. There are fewer examples, and the problems are harder and come without answers.

So the new books enjoy colour and a more attractive layout, come with many examples, and with answers, so the students can check whether they have done the problems correctly. It is also often possible to buy an additional book, which contains worked solutions to the problems.

This development has been caused partially by the success of the Schaum Outline books. (I take this opportunity of pointing out our indebtedness to Dan Schaum, a man who had a great reverence for learning, the author of the first Schaum Outline series, and the founder of the whole series.) The Schaum Outline series teaches mathematics by first giving a very brief sketch of the theory, often with the statements of theorems without proofs, and then giving a large number of problems with their solutions.

Many academics were dismissive of these books, because they felt the students should not take the easy way out, but struggle to find the solutions to problems themselves. There is of course much point in that, provided you can find your way to a solution to a reasonable number of problems in a reasonable time. Otherwise you are spending hours of frustration and disappointment getting nowhere. At least with the solutions you can do something.

In general it is important that students who are urged to work harder should do something which is productive. Instead of preaching at the students, they should be given some practical and useful task which they can carry out. If the only thing students can do is painfully follow the details of the solutions in the crib, it is a lot better than just staring hopelessly at the problem. I have been quite impressed too that many students do seem to learn a lot in this way. If at the end they still can't do problems on their own, at least they have learnt something.

Nevertheless, the better students do need to be encouraged to try to persevere and answer the questions without the answers.

There is a type of text-book, quite common in Russia, which is similar to the Schaum Outline series in that the book concentrates on problems, but unlike the Schaum Outline series, does not include the solutions. Thus the books consist of a series of sections each of which begins with an example or a definition, and is then followed by a series of questions, which should be done in order. The sections are skilfully organised so that skipping a question would make the next problem too difficult, but if one does the questions in sequence, the transition from the one to the other is reasonably straightforward. In this way one systematically goes through the subject matter. The advantage is that the student is very active, but needs only consider a narrow subject area at a time, thus learning with gradual increments. An example of such a book is Burn [1992].

These books tend to have reasonably difficult problems. Another approach is taken by the programmed learning type books, where the questions are deliberately made easy, so that the reader is encouraged by getting the right answer.

In §5.2 I mentioned the skill of skimming through a mathematical book, to get a general idea rather than learning the subject matter in detail. The trouble is that most mathematical books are not written with this in mind. It is time that they were! It only needs a little care, the more restricted use of symbols, more introductions and summaries, and more guidance throughout the book. The habit of putting in the minimum of information and explanation does not make for easy reading.

However, even worse is a tendency seen in some North American texts of waffling. One needs concise, clearly written texts, which are prepared to help the reader.

Perhaps the biggest problem with set books is that they must be universal books: they have to be comprehensive, so that most courses can be based on them. It is in this respect that the lecture-based system is superior, for it is quite possible to give a shorter, more direct course if you can yourself decide which topics to cover, and how carefully.

As mentioned above, the new text-books come with clearer and more attractive lay-outs, with colour and better diagrams. This can be taken even further, as is evidenced by Giancoli [1998]. All the techniques of a hard-selling holiday brochure are in evidence here. This is not a mathematics book, but a physics book, but it does indicate how far one can take these methods, since it is not difficult to see that one could do something similar for mathematics.

The front cover of Giancoli's book features a spectacular view of a winter skier, a wonderful feeling of movement, with snow everywhere, flying up in scurries. Throughout the book there are magnificent colour photographs. Diagrams are enlivened with colour. There are colour head-lines and shading. In addition there are a host of interesting current applications, such as lenses with non-reflecting coatings, the Hubble telescope, forces used in straightening teeth, photocopiers, air bags, bungee jumpers. Even the standard diagrams show the benefit of an artist's hand. For example, a simple diagram to calculate the height of a building, nothing more or less than a triangle, is enlivened by a stylised building, with colour shading that adds brightness, even glamour.

This book is colourful, enticing, attractive, seductive. Does it help teach physics? To some extent, yes. It is exciting. It seems very relevant and useful. But there are also disadvantages. For instance, there is so much to read and tempt one that one fails to get down to business. Thus for a book of this size, there is little deep material. This is also not a book that can be written quickly. It has clearly drawn upon the efforts and skills of many talented people, artists, lay-out experts, and ideas and examples from physicists in addition to the author. This is a book that requires enormous commitment and resources to produce. Yet some of the ideas and techniques could be introduced in a limited way in other books. It is a useful book to keep in mind.

Finally I would like to suggest authors of text-books add more information of their experiences using their own books. It would be useful to know something about the ability and pre-knowledge of their students, whether the book is regarded as difficult or reasonable or easy, what percentage pass, how much of the book is studied, whether their course is as formal as the book, etc.. The number of hours of lectures and tutorials per week and how many weeks of intstruction, and also the students' other commitments, would also be useful.

10.4 Using computers in the teaching of mathematics

I have already commented in §5.4 on modern technological methods of teaching, which are currently of great interest but are not yet in wide use. Nevertheless the computer itself has become an important part of mathematical education. The department must decide what part computers will play in its courses even if they are not used to supply the new computer based methods of teaching.

Computers come into play in three main ways. Firstly the use of the computer extends the value and application of mathematics enormously, in that calculations which previously could only be done theoretically, can now be done practically. Next we can use the computer to calculate a series of examples which give us a concrete understanding of theoretical problems, and can also help us to formulate and explore new questions. And finally we can use the computer to provide us with teaching devices superior to books so as to teach mathematics better.

Because computers now make it possible to really use mathematics it is important for students to learn how to program and to learn to use different programs like Matlab, Maple and Mathematica. They need also to learn how to apply the mathematics in these different formats. It is also crucial to be constantly on the look-out for ways of checking the results obtained with the program. For instance, it should be possible to see whether the result is reasonable, by making a rough estimate, or to test the result by seeing whether the program gives the right answer for a similar problem where the answer is known by some other method.

If one is going to use the computer to gain a feeling for abstract problems one needs to get used to programming. One needs to spend much time gaining confidence in the use of a particular computer program, so that using it is almost second nature. That is why I am in favour of beginning the first year with an introduction to Maple or Mathematica or to some spreadsheet program, like Excel. Students should have the possibility of verifying theoretical results by having access to these programs. They can be used to draw graphs, calculate integrals etc., thus checking up and reinforcing the theoretical studies. It should be as natural to use these programs as it is to use pen and paper. With the large increase in the number of home computers this is rapidly becoming a real possibility. If one needs to rely mainly on the university's computers, the demand can easily exceed supply.

For instance, it has been suggested that the spread-sheet and the graph drawing capabilities of a hand calculator can lead to a clearer understanding of the idea of a function. But this can only be the case if one feels confident in handling these programs.

The disadvantage of learning to use such computing programs is that it takes time to learn, time which could perhaps be better spent on learning mathematics. It is because of this that a spread-sheet program is useful, because it requires very little to learn to program, and it is especially useful for recurrence relations. Two very simple but rewarding exercises on a spread sheet are Euclid's algorithm and approximating a square root. Another is convincing oneself that $\sin(x)/x$ tends to 1 as x tends to 0.

Another way in which technology could be used is in helping us to learn. One can use the computer as a super book, in which interesting visual effects can be achieved. For instance, one advantage that a lecturer has over a book is that the lecturer can point to various parts of his equations and show how they should be interpreted, but this too can be done with computer programs. Then one could request and be given various exercises, and the user could state the degree of difficulty required, with the solutions being marked instantly. Another way is that the computer could be a vehicle for a programmed learning course. Courses of this type take much time and effort to prepare, and consequently can lead to courses which must remain unaltered for many years so that they are economically viable.

10.5 The overhead projector and the blackboard

The blackboard (now often replaced by a white-board) is a very satisfactory way of presenting material, because one can show the students how one works out the mathematics on the spot. It is like working with pen and paper, the standard tools of the mathematician, and the tools the student has at his disposal. You are showing how to do such calculations.

Of course, one can use the overhead projector in this way, writing down calculations in the same way one would write them on the blackboard. There are two disadvantages. The first is that one tends to be blinded by the glare of the lights. The second is that only a small amount of material can be displayed at any one time. One can of course use two overhead projectors simultaneously, which does improve the amount of material one can display simultaneously. But a large expanse of blackboards is even better. The single overhead projector is not satisfactory as a direct substitute for the blackboard. Then too many people fail to use the overhead projector in a sensible way. They tend to come with prepared sheets, scores of them, and quickly and dextrously put them on and off display, before you have a chance to read them. Often these overheads are a miracle of close packing, with vast quantities of information compressed on them, difficult to see and read.

On the other hand, there are some cases where the overhead projector can be used to advantage. It is good for pictures, for quotations, or cartoons. Although you can provide beautiful pictures, and even add movement with overhead projectors, what one really aims at is showing how students themselves can produce their own pictures, models and diagrams in a simple but effective way; thus I suggest you avoid the showy, attractive professional methods most of the time.

So on the whole, the overhead projector is an inferior method of presenting mathematics. What is needed are large moveable blackboards, preferably stacked in threes, one in front of the other.

There are three main dangers in using the blackboard or white-board. The first is that one blocks the students' view. So it is sensible to move aside to ensure the writing is visible. One should also avoid speaking to the blackboard or whiteboard. Finally one should not march backwards and forwards, which tends to mesmerise students instead of instructing them.

10.6 Different methods of teaching

There are now a large number of different alternative methods of teaching which are accepted and regarded as normal: lecture courses, small group teaching, programmed learning, the Keller method, problem based learning, seminar based courses, reading a book and so on.

Ideally one should select the method of teaching which best promotes one's aims. It would then be quite reasonable to use different methods of teaching for different purposes. For instance, teaching in groups helps new students adapt to the university and find friends. The Keller system can be used for a very mixed group of students, to raise them all to the same common background and knowledge. Programmed Learning could be used to give drill for certain essential skills. Lectures can be used to give an introduction and overview, etc. A brief list of various methods of teaching with advantages and disadvantages now follows.

I make no pretence to comprehensiveness, the possibilities being so large. Nor does the choice of systems I give indicate that they are better than others; it is simply that the choice is very wide, and I do not have experience of all these systems. In particular, there is much activity and interest in computer aided learning, teaching mathematics by Socratic discussion, and teaching by writing. I refer the reader to Schoenfeld [1990] and its further references. The discussions in Burn [1998] are also very useful.

Method 1: Lectures type A
In my present university, Mälardalen Univeristy, a typical course held during a 8 week period would consist of 28 hours of lectures to 60 students, 18 hours of a problems class with all 60 students in which problems are solved by the lecturer on the whiteboard, and 16 hours of tutorial work with students divided into two groups of 30 each. An example of such a course is Linear Algebra, based on the book Norman [1995].

Advantages: This provides an efficient, disciplined method of learning. Ideas can be presented in alternative, less rigid ways than in a text-book. Lectures can be tailored to suit the course and the students. Where there is no suitable book, this provides and easy way of developing the material and the course, and many a book has arisen in this fashion. It is easier to ensure that students are not subject to an excessive load since one can see how they react to the material and one can slow down or speed up if need be.

It is easier to provide an overall view than in a text-book, to emphasise various ideas, and give some of the tricks of the trade. It provides an opportunity to see how a mathematician solves problems, i.e. to see the scaffolding and not only the final product.

If the students re-work the lecture after each occasion they have a systematic way of studying and revising. Increases in the number of students can be accommodated quite easily.

Disadvantages: The students must all have roughly the same pre-knowledge and ability. Students need rigid self-discipline. Thus they must re-work the lectures and must realise that subsequent lectures can only be understood if the previous lectures have been properly absorbed. If the lectures are based on a book, the student must read ahead.

If the student loses track of the argument much of the rest of the lecture is valueless. Students can fall into the habit of simply taking notes, thus deluding themselves into thinking they are doing worthwhile work.

There is little contact between student and lecturer. The student has no check on his work, and not much incentive to work. It is difficult for students to ask questions. The tutorial and the student's own work, vital parts of the instruction, may be regarded as unimportant compared with the lectures.

Hint: Students find concentrating for more than 20 minutes difficult. Therefore arrange breaks every so often. For instance, students can be given 5 minutes to do a calculation instead of the lecturer doing the calculation. See §11.12.

See §6.8 for advice to students on how to get the best out of lectures.

Method 2: Lectures type B

Much the same as Method 1, the difference being only in the number of students attending the course, which would be half as much, i.e. 30 students in all. Thus typically there would 28 hours of lectures, problems classes of 18 hours and tutorial of 16 hours, with all 30 students at each session.

Advantages: The advantages are the same as for method 1. Also the smaller group leads to a more personal and friendlier atmosphere. The possibility for students to ask questions is greatly increased, and if the number of students is below 20 or 25, it is possible at times for the lecture to become a dialogue.

Disadvantages: The disadvantages are the same as for method 1. But there is a more personal atmosphere and asking questions is easier.

Method 3 Keller method

Students study on their own, usually reading a book, and then are tested before being allowed to proceed. Directions for studying, problems, tests etc. are given in hand-outs called units. Students meet their tutor for half an hour once a week over an 8 week study period.

Advantages: There is good interaction and feedback. Since the student is seen on a one-to-one basis, it is much more personal. The student is constantly checked, and only allowed to continue if the material is sufficiently well understood. Thus this is a good method for students of widely differing skills and knowledge. There is more discussion. Students have to learn how to read a book. On the whole this seems best suited to mastery of routine material.

Disadvantages: This is a lonely method of studying. The rate of learning can be slow. It is very much a directed system of learning. Students have to work more hours to learn the material. The book assumes central importance, and it is difficult to find suitable books. Some tutors and students just do not get on well together. Tutors can find repeating the same things boring. Continuous testing imposes a strain on the students. This method is expensive in staff and student time, and takes considerable time to set up.

Method 4: Teaching in groups of four students

Groups of four or five students study together in a class-room. They have two sessions per week each of four hours. The teacher shares 8 hours per week among six groups, thus handling 24 to 25 students. When the teacher is with a group there are discussions and tests, and instructions what to do next. The students have to hand in problems for correction, sometimes on an individual basis, and at other times as a group. Oral answers are also required and are evaluated. The total period of study is 8 weeks.

Advantages: Students help one another which benefits both those who explain and those who listen. One learns how to work in a group. This is a friendly and sociable way of learning. Some groups interact extremely well, and forge alliances which contribute to their studies in other courses. Much of the advantages of method 3 apply also.

Disadvantages: It can be difficult to arrange groups with students of equal ability, and a group of students who differ too sharply in ability does not function well. Some groups do not work well together for reasons of personal chemistry. There is no well thought out systematic exposition, the tutor has to answer the problems when and how they come, and can have difficulty in getting a well thought out answer on the run. Students can learn inferior, even wrong methods from one another. Better students resent having to help weaker students all the time. Weak students studying with strong students can have their weakness emphasised, and their feeling of inferiority reinforced. The disadvantages of method 3 apply, but there is no feeling of isolation.

Students can not organise their time themselves but must keep in step from week to week. Very much a directed form of study with strong tutor control. If the number of students increases say to 30 the system begins to break down, there being insufficient time for staff/student interaction.

Hints: This method works best in conjunction with short lectures (which include guidance, a short introduction and a brief summary) to all 25 students simultaneously. Avoid examining the students too harshly. Sufficient time must be allocated to each group.

Method 5: Guided reading

Each week the student reads parts of a set book. The tutor and student meet for half an hour each week for some 16 weeks. The student is given advice, suggestions, and the subject matter is discussed.

Advantages: This provides strong motivation, and can provide a pleasant and personal interactive environment. Errors can be easily detected and corrected. It is easy to provide an overview, and introduction to the following parts of the book, and additional ideas.

Disadvantages: The student works with no interaction with other students. The student must be very skilled and capable of reading and studying alone and must be strongly motivated. There is no direction other than the weekly discussion. This method is expensive in staff and student time. The tutor and student may not get on well together.

Method 6: Project.

A typical example say would be to discuss the solution of a polynomial equation of degree 3. Or else to discuss symmetry groups. The student would have to read from a variety of sources, and then write a report. Examples of more extensive projects are given in Appendix H.

Advantages: Similar to method 5. Often students feel tremendous motivation and inspiration because they are doing a project they themselves have chosen.

Disadvantages: Similar to 5 above.

Hint: It is an advantage if students can given a choice of topic, rather than be forced to work on a project that is assigned to them.

Method 7: Oxbridge method

Lectures are given to large groups. Two tutorials, each of an hour, are shared between 2 students with a single tutor per week.

Advantages: Regarded by many as the ideal combination, the Rolls Royce of teaching. One gets all the advantages of the lecturing method plus the advantages of a more personal system. The student is subject to strong pressure to work, receives detailed advice, help, correction, and encouragement.

Disadvantages: Very strongly dependent on the tutor. In many ways a directed method of teaching with compulsion (the student is expected to hand in work regularly). Expensive in staff time

Method 8: A constructivist method

This is a method which has been used at Purdue University. Attendance at the time-tabled sessions is compulsory, and the course is organised co-operatively. A cycle of study begins with laboratory work in which the student carries out computer tasks to help each individual's mental construction of concepts. These laboratories are then followed by a class involving a modified Socratic approach . Finally the students do problems. The design of the course is guided by careful observation of and interviews with students.

Advantages: The main claimed advantage of this method of teaching is superior learning. Through the laboratories students are encouraged to think out many results themselves. The use of the computer makes the studies much more concrete. The Socratic discussions ensure the students think actively.

Disadvantages: Sufficient computers are required, preferably with a simplified programming language. Special text books are required. Teachers and students must learn a new way of instruction. Students tend to work 2 or 3 times the number of hours of an ordinary course. (Some regard this as an advantage.) Teachers have to work harder. The original setting up of the course requires a huge expenditure of staff time.

[Since I have no personal experience of this system the above remarks should be regarded with caution, and for information one should consult the appendix by Ed Dubinsky in Krantz [1999] and the further references given there.]

Method 9: Programmed learning

This type of teaching arose as a consequence of Skinner's success in teaching pigeons to carry out complicated tasks. This was done by splitting the task up into a large number of simpler parts. Each of these parts was taught by rewarding the bird whenever it did the right thing. And the tasks were as far as possible made relatively easy, so that the bird was rewarded frequently. Typically the reward was food.

In the case of programmed learning for human beings, the reward is getting the correct solution to a set task. The work to be learnt is broken up into small components, and there are questions to be answered following each new section of material. The questions which are asked are made fairly simple on purpose, so that getting the right answer is not too difficult. Whenever students gets the right answer and check and find out they are right, they feel, rightly, a glow of satisfaction, which keeps them eager to continue to study.

At one stage there was a vogue of books with this technique of teaching. A variant occurred in some of these books where if you had got most of the answers right, you were routed through the material in another way, so you did not need to read the whole book and answer all the questions, and so could go through the course more quickly. The same technique is now available on computer, where it is possible to be more flexible.

Advantages: Constant repetition ensures thorough learning is achieved. Careful testing ensures students have no gaps in their knowledge. Best for learning facts.

Disadvantages: It takes considerable time and effort to set up, and can turn out to be tedious. Learning takes place slowly. This is very much a directed method of teaching, and as such should be used with caution.

Method 10: Problem based learning

Problem based learning is a method based on studying a particular problem, the solution of which requires knowledge of number of subjects to solve. So that, for instance, instead of studying a course on Statistics, and one on Calculus, and one on Algebra, and then combining these to solve a problem, the students start with the problem, and then study those parts of Statistics,

Calculus and Algebra which are needed to solve the problem. The students are often organised into teams.

Advantages: This method is more like the work that one does once one has left the university. One learns to work in a team. This method provides a motive for studying. Makes sense for students who have learnt a fair amount of mathematics already.

Disadvantages: Not good for beginning students. Expensive in both staff and student time. Fails to give a thorough grounding in the subjects studied.

Method 11: Seminar

Each student presents a seminar of an hour. Each student in addition is expected to ask questions and to comment on the seminars presented by the others. Two staff members assess the grade. Students are allowed to seek help and advice in preparing the seminar.

Advantages: One learns to give talks. Since one chooses one's own topic, students are very well motivated. The topic can be arranged to suit the student's abilities.

Disadvantages: One works extremely hard for one's own talk and then takes it rather easy for the others. Can be very stressful. It is hard to grade the seminar objectively.

There are many other alternative methods but these give some idea of the variety.

It is also worthwhile assessing the number of hours of work required both by students and staff. For instance, for Method 1, the number of lectures is 28. I estimate that it requires 28 hours for the lecturer to go through the material revising, and another 28 hours to prepare adequately to give the lectures. That is 84 hours altogether, plus a total of 50 hours spent on tutorials. Thus the lecturer spends 134 hours per 60 students, which works out to 2.2 hours per student. Each student spend 28 hours in lectures and 34 hours in tutorial work. If the student matches every class hour with an hour of own work that gives an extra 62 hours. The students I interviewed suggested that they studied some 100 hours extra on their own.

Method	Lecturer hours /student	Student hours/course
1 Lecturing to 60	2.2	162
2 Lecturing to 30	4.5	162
3 Keller system	5.1[9]	214

Fig. 10.1 Estimates of student and lecturer hours for various methods

It can be of interest to do these calculations and see what results one obtains. The results depend on a variety of variables, the type of student, the type of course, the particular organisation etc., and thus Fig. 10.1 for my students and institution is unlikely to apply to your own institution. Consequently I have only compared three methods.

It is also of value to discuss these various systems using the rules of teaching in Chapter 7.

10.7 Indirect methods of teaching

There are indirect methods of teaching, where one absorbs ideas and information, almost like a process of osmosis. One learns about the matter very gradually, by hearing about it, even if one does not really understand what one hears at the time. So one should deliberately include material which the student can not possibly really understand just at that moment. Some of these ideas will gradually gel in the mind, and will form the basis of later understanding.

For instance, general lectures on a specialised topic may be worth arranging, even though they are above the level of immediate understanding.

There are all sorts of interesting lectures one can give, which provide a flavour of research mathematics, although on a lower level.

In England the BBC organise each year a series of lectures called "The Christmas Lectures". Mainly meant for children, they provide a fascinating view of some important branch of science. It is feasible for the academic staff to provide some general lecture on mathematics in a similar style, although it woiuld be very difficult to reach the same high standard.

An important source of incidental learning comes from the library. All students, good and bad, will learn something from the library, if there is a

[9] Excludes the design and preparation of the course.

sufficient variety of interesting books available. So in building up the library, one must think of providing books, periodicals and videos which will enlarge students' capacities and interests.

For instance, the books of Martin Gardner, Keith Devlin, and Ian Stewart should be available. Periodicals are also important. The Scientific American, the Mathematics Intelligencer, the Mathematical Gazette, The American Mathematical Monthly, Mathematical Spectrum, and the College Mathematics Journal should be available. See also Appendix B for a list of video sources.

Unfortunately providing a good library is extremely expensive, and it is hard to justify books which will be used only occasionally. And yet without a library no university can really exist. I have been fortunate in most of the universities where I worked that there were very good libraries. In the present university, Mälardalen University, there is a very modest library. This is not surprising, since it is a new university. The American Mathematical Society has a suggested list of suitable books to equip a new library, but most librarians will blanch when they see it. So there is nothing much other than gritting one's teeth, and asking, again and again, year by year, for the provision of library books. Over a period of five years, there will be a significant improvement.

A newspaper on the internet is theoretically very easy to organise, with references to interesting articles, and some interesting problems. Members of staff could be asked to send in contributions. It is best to appoint a member of staff as editor, and provide time off from teaching duties in order to organise the newspaper.

10.8 Conclusions

With this chapter we have concluded our discussion of matters which the department needs to co-operate on as a whole, and in the next two chapters we turn our attention to what each individual can do to organise and improve teaching.

PART IV THE INDIVIDUAL LECTURER

Chapter 11 Lecturer's Approach

Chapter 12 Some Practical Points

Chapter 13 Assessment of Teaching

11. Lecturer's Approach

Preview

In this chapter and the next we will consider what the individual lecturer can do to improve teaching. Here we are mainly concerned with the lecturer's general approach and attitudes, and the way in which the teaching is arranged. It is important to motivate students, and one therefore needs to take into account their interests. Introductions, repetition and explanation are important in teaching.

There are many things which can be done to improve any teaching, but each of these has a limited value and takes a fair amount of time, and the biggest problem is to decide which idea or method to use at the expense of which other idea or method.

11.1 Introductory remarks

In Mayhew [1990] the opinion is expressed that it is impractical to expect a lecturer to spend too much time working through the literature on teaching in order to improve. After all, that and planning to produce courses in such detail as many of these methods advocate would be more than a full time job, let alone the teaching and the research and the administration that still has to be done.

There are however a few books which definitely should be looked at. Most of all I would recommend the books Polya [1965,1977,1985]. Krantz [1999] is sensible and useful and is written by a mathematician for mathematicians. It includes a number of interesting appendices written by mathematics teachers with differing views, and from them one gets an idea of how large the variations in thought on the subject are. Schoenfeld [1990], the Mathematical Association of America source book for the teaching of mathematics, is of sound practical value. Burn [1997] is filled with ideas, suggestions and questions.

Tall [1991] is a more theoretical book, interested in mathematical thinking and creativity, proof, definitions, the role of the computer and much more, and is in a different direction to the present book. Griffiths [1974] concentrates on school mathematics but has many interesting views which are of relevance to university teaching. Indeed, since most research into teaching involves teaching at school level, it is often of value to look at books on teaching at school, such as Glatthorn [1993]. Mayhew [1991] recommends the general books McKeachie [1978] and Eble [1976] as the most practical way of improving and maintaining standards of university teaching. Ramsden [1992], and Beard [1984] are also concerned with teaching as a whole, and are not directed to science or mathematics.

There is a small pamphlet by the brothers Clark [1959] which is worth looking at, just as it is worth remembering their advice: "...lift up your head, look at your class, speak out boldly, teach them what they need to know, watch them continually to see how they are taking it, and all the rest will be added unto you."

Books like Parsons [1976] and Northedge [1997] give advice to students on how they should study, and if you think of it from the students' point of view, it will help you design your courses.

The following practical suggestions are motivated by the rules of teaching in Chapter Seven.

11.2 The lecturer's approach

The most important attitude for a lecturer to have is to be on the students' side, to feel that it is important for them to succeed. They may very well have the "wrong" motives for study, e.g. simply so as to pass. But they are entitled to their motives. It is up to the lecturer to teach them something of value. By giving them to understand that the lecturer wants them to succeed, there is very much more chance of their learning. It is important to illustrate to them how the present course will help them achieve their needs.

Then too you must be encouraging. For instance, one can almost always say quite truthfully after a couple of weeks have gone by, "I am beginning to notice an improvement. You probably can't see it, but I can. So keep up your efforts." One need not pretend, for in fact, these improvements are often quite considerable. But the students can't notice it, for no sooner have they understood one part of the course, than they are onto another part of the course, so that they are continually struggling with apparently no success, and constantly depressed. Tell the students to look back and see what they have accomplished.

Then too the lecturer can not afford to lose his/her temper.

As a lecturer you hold a great deal of power. You can abuse it. You can come to lectures late with impunity. You can fail to mark the tests sufficiently quickly. You can mark somebody down. You can prepare poorly. You can arrange tests which are too difficult. However, you must be your own strictest critic, and ensure that you do an immaculate job.

Now all of this is relatively easy to do with very good students, a little harder with weaker but at least serious hard-working students, and very difficult to do with reluctant students. Things are worse in North America, where sometimes students and staff dislike one another deeply; the staff because the students are noisy and uninterested, and the students because they have been forced to take a course for which they have no aptitude and they find it beyond their abilities. Nevertheless, this is part of what the job is. Somehow enthusing the reluctant students.

If at all possible prepare much of the routine material in advance. Thus have the examinations and tests prepared before the course begins. This not only prepares you for the course, it also ensures that you are unlikely to suddenly have an overload when you have to mark the test and prepare the exam paper. In other words, make sure that you have some leeway so that everything can be done on time.

If you carefully think through the commitments of the term in advance you are unlikely to be caught off balance and give a poorly prepared lecture or be late with your marking. Students like to have the results of their tests as soon as possible. You can't do this if you are busy preparing the next lecture or examination. If you can't mark the work quickly, do at least tell the students when they can expect the results, and ensure that they are out on the promised day or earlier.

A lecturer must be enthusiastic. If you are a mathematician you must be able to see the timeless beauty in the subject you are teaching, and even if you have taught the chain rule for differentiation a hundred times, you must still be aware of its continued importance and interest.

Being enthusiastic means that you must also prepare your lectures properly. It does not suffice to glance through last year's lecture notes fifteen minute before the lecture. (Surprisingly many do this.) It is important to work through each lecture carefully, and it is a good idea to change the courses each year. You will always find things that you could do better from one year to the next. You should prepare so well that you should be able to give the lecture with only an occasional glance at your notes. The need for this is that you must keep looking at the students, because only in that way can you assess whether they are following the argument, or whether you need to modify your delivery, either by adding more and better explanations, or else by going on to the next topic, when you have made your point. Looking at the students also makes the delivery more personal and more interesting.

Whether one will look at the students in one's career as a lecturer probably depends most on the very first lecture. Nothing before in one's experience has led one to expect the sight that one faces. During a lecture students seem to change completely. A perfectly normal person suddenly takes on a look of incomparable boredom. Some students may be sleeping or doing crossword puzzles, some couples may snuggle up to one another. One can understand how a lecturer can take one terrified look at this sight and with a shudder, resolve never again to look at a class. This is a serious error. The class gives many valuable indications of how your explanation is working. If they do look puzzled or bored, you will be spurred on to add to your explanation. If you do not look at them, you lose this valuable information.

Do not forget the first rule of teaching, namely that the course must be at the right level. The diagnostic test (see §8.2) will help you gage this. Another useful technique is to make a list of appropriate topics and ask the students which of these they are familiar with.

Inevitably students are used to TV presentations. An ordinary lecturer can not hope to produce anything as impressive as a talk on TV, where there is a host of people who have worked on it, a script writer and a director, a speaker who has spent much time in practice, beautiful pictures, and stirring music. So students are inevitably a little disappointed in comparing your performance with that of TV. Also, they are not used to concentrating during a TV talk. They may very well be used to preparing a cup of tea, or sorting through their mail, or getting dressed, and they bring these habits of watching TV to the lecture theatre.

I was once told the difference between a good carpenter and an indifferent one. The good carpenter, after finishing a job, sharpens his tools before putting them away. This is a useful point to remember. When you have finished your lecture course, think about it carefully, make notes about your experiences and the conclusions you have drawn, improve the course, file away your notes carefully, so that the course is ready for use next year. You can also record your experience in a uniform fashion from one year to the next, and in §13.2 I suggest a way of doing so.

11.3 Put in the construction lines

Artists often use construction lines to help drawing. For instance, to ensure that they have the right perspective, they provide a vanishing point and draw lines from the object to the vanishing point. Afterwards they rub out the construction lines. As far as the finished picture is concerned, they should not be there. But as far as the aspiring artist is concerned, they are most helpful.

The same thing applies to mathematics teaching. The final perfected argument, without the construction lines, appears in the text-book. The students do not need you to basically repeat the text-book. They need a rougher version of the work, so that they can see how it is possible to construct the final perfected version. They need to see how you derive the results, so that they themselves can do so.

11.4 Introduce, summarise, recapitulate, revise

My advice is that you should introduce, summarise, recapitulate, revise. You can not do these too often. Try diagrams, showing the relationship of the newer material to the older, and how you used the various definitions in your material. (Useful ways of drawing diagrams are explained in Nelms [1964], Buzan [1989] and Northedge [1997].) Having introduced new material, recapitulate the old, and show how it relates to the new.

I always begin my courses with an introduction to the whole course. I go through some of the vital results or theorems in some form or another. The idea is to do so in language the students can understand, so the introduction is necessarily somewhat confined. This type of introduction is different to that sometimes given in books, in which the aim is to tell the expert reader, the lecturer on the course, how the book has been designed, and is of no value to the average student who does not understand the terms. The idea here is instead to try to explain the subject matter before one has learnt the definitions and main theorems.

For example, here is a suggested introduction to an elementary calculus course:

One of our aims is to understand the behaviour of a function, and to do this it is invaluable to sketch the graph of the function. By a function I mean for example,

$$f(x) = \sin(x) + 2x^2, \text{ or } g(x) = x^2 + 1/(2x + 3).$$

A fundamental tool for the sketching of a graph is the slope of the tangent to the curve. Knowing the slope of the tangent tells us for instance whether the function is going up or going down. The slope of the tangent leads to the idea of a derivative of a function, a crucial concept.

In fact there are three main characters in the course: The derivative of a function, the integral of a function, and a primitive to a function. These characters play the main roles in our story. They crop up everywhere.

The derivative of a function $f(x)$ is denoted by $f'(x)$. It is, as I have just mentioned, the slope of the tangent. This is a rough definition which will suit us for the moment; a precise definition will be given during the course.

In Fig. 11.1 we have the graph of a function, and a tangent. It is the slope of this tangent which is equal to the derivative of the function at that point.

Knowing what the derivative is of great help in sketching the graph of the function. Fig. 11.2 enables us to say using the concept of derivative whether our function is increasing or decreasing, i.e. going up or down.

With these rules it is considerably easier to sketch the graph of a function, which is as mentioned before, a useful way of understanding its behaviour.

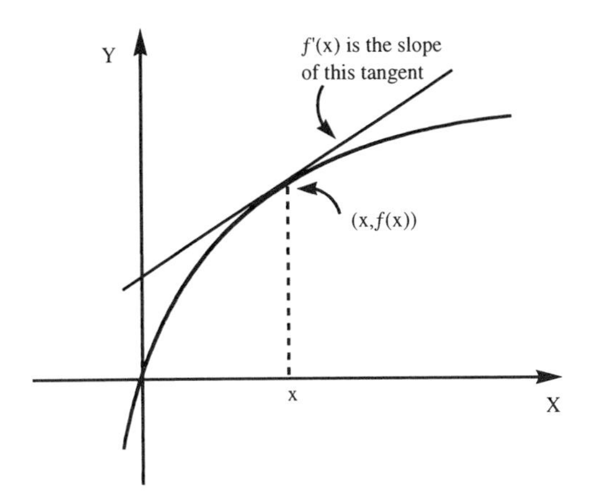

Fig. 11.1 The derivative of $f(x)$

Derivatives can be used to interpret such important physical concepts as speed, and acceleration, and are used in Newton's rules of motion to provide us with a proper understanding of planetary motion, for instance, giving us an explanation of why the moon goes around the earth, or enabling us to predict the path of a satellite.

Sign of f′(x)	**f(x)**
Positive	Increasing
Negative	Decreasing
Zero	Maximum, minimum, or inflection point.

Fig. 11.2 What the derivative tells us about the function

The next main concept is that of an integral of a function between certain limits. The intuitive definition of the integral between the two points a and b (which we denote by the symbol A(a,b) for the moment), is the area under the curve, as shown in Fig. 11.3.

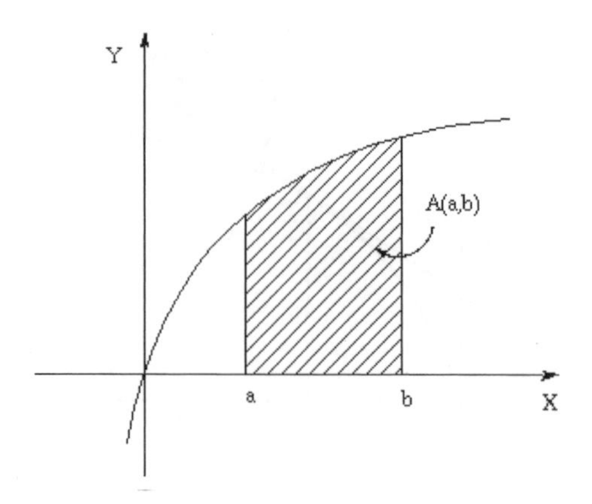

Fig. 11.3 The integral of a function

This concept one might think is only of use in calculating areas, but nothing is further from the truth, because it can calculate all sorts of valuable and important quantities. Volumes, density of an object, the work done when moving a particle from one place to another, electric charge, and many more interesting and useful physical and mathematical quantities. Basically one can think of it as an infinite sum, and what this means will be made clear later.

The third main character, a primitive function $F(x)$ of $f(x)$ is simply the opposite of the derivative, i.e. a function such that its derivative $F'(x)$ is equal to $f(x)$.

One can almost describe it as a different way of saying $F'(x) = f(x)$, that is, we say equivalently that $F(x)$ is a primitive function of $f(x)$. It is just another form of words. And yet this concept is of great importance.

There is an important connection between the integral and the primitive function, the so-called fundamental theorem of Calculus, without which Calculus would never have got off the ground. It is this which makes it possible to calculate integrals.

But to go back to derivatives again. It is quite hard to find derivatives from the definition alone. So it is useful to have rules which enable one to calculate the derivative of a sum or product or quotient if one knows the derivatives of the constituent functions.

A useful theorem enables us to find local maximum or minimum values, for these occur when the derivatives are 0.

Compared with finding derivatives, finding primitive functions is very difficult, but there are a number of useful techniques.

Everybody loves a special type of function, namely the polynomial. It is easy to calculate, to find its derivative, and also to find a primitive function. That is why the theorem called Taylor's theorem is so valuable. It expresses a function with derivatives as a polynomial, called its Taylor polynomial, plus an extra term, a so-called remainder term. When the remainder term is small, the Taylor polynomial becomes a very useful approximation to a function. For instance,

sin(x) is approximately $x - x^3/6 + x^5/120$.

Both derivatives and integrals depend crucially on the idea of a limit, and this final important concept will be developed later in the course. Logically it comes first, but since its importance becomes clearer once the other topics are developed, we will delay its introduction till the end of the course.

So this course is concerned with finding derivatives and integrals of various functions, and these enable us to understand functions extremely well, to find maximum and minimum values, to calculate volumes, work done, lengths of curves, areas. As such it constitutes the fundamental mathematics required for much physics. The main concepts are limit, derivative, primitive function and integral.

The reader will note that there is no clear or accurate description involved in this summary, but at least one has some idea of where one is going. The reader will also note a large amount of redundancy and repetition[10], important in a lecture, unnecessary in a book.

Here is a second example, for introducing point-set topology:

Long, long, ago there lived in a certain country a young mathematician. He studied hard, and all went well for him, and after a while he became chief of the Government Analysis Department. He was an expert in finding limits, in checking for continuity, and in applying the intermediate value theorem.

One day, when he was out to lunch, a thief broke into the office, and stole the young mathematician's ruler, or more formally, his distance function. No longer could he measure ε or δ or decide whether two points were close to one another. His life was in ruins.

In despair he suddenly thought of something. In all the years he had been at work, he had in his spare time, collected open intervals such as (a,b); it was a hobby of his, he had all these intervals lying around the office, in no

[10] Even more elaboration would be needed in the actual lecture.

particular order, and alas, not even labelled. Perhaps he could find a way of using them instead of his ruler.

And he succeeded. He found that with his open intervals he could define such ideas as limits and continuity, and talk about points that were infinitesimally close to a given set, and all without using his distance function, which he could not possibly have managed before it was stolen.

Other theorems like the fact that a continuous function is bounded on a closed interval and the intermediate value theorem had to be explained in a radical new way, which nevertheless brought a greater naturalness and understanding to these ideas. The intermediate value theorem turned out to have something to do with the geometric concept of connectedness, for instance.

Having discovered this, our young mathematician realised that there were all sorts of structures, which he could define by means of sets which he called open sets. These structures in time began to be called topologies. Some of them came from new distance functions, and some could not possibly have come from a distance function. Thus began the study of topology

Another good way of introducing a course is to give a list of problems that will be solved in the course.

The value of these introductions is one gives direction to the material, and see the subject as a whole, so that it has more meaning. In this way, I am emphasising the Gestalt theory of learning, that one learns as a whole rather than in small bits, thus giving a wide-angle view, rather than a close-up view.

For instance, to understand the town Västerås in which I now live, is relatively easy. One could begin by pointing out that the river cuts it in two from north to south, while the motorway cuts it in two from east to west. Eastwards 110 kms brings one to Stockholm while westwards 400 kms brings one to Oslo. The main road is Vasagatan which runs from north to south. In the south is the railway station; further south brings one to the lake. Walking fifteen minutes north from the station brings one past the town hall, the cathedral, and then to the university.

Or alternatively one can get the student to crawl on his hands and knees, studying in detail with the aid of a magnifying glass, each paving stone. I exaggerate of course, but it is the latter approach that we seem to use in much mathematics teaching.

Despite my preference for the overall view, the splitting up of complex ideas into small sections, each of which should be mastered a bit at a time, and then assembled as a whole, can often be a very good method of teaching. Another point is of course, such is the variety of minds, that there are certainly students who prefer the slow, detailed bit by bit development of a subject, rather than trying to see it as a whole, for all their studying.

Nevertheless, having mastered this detailed approach it is important to try to get the overall view. Continually giving introductions, summaries and reviews enables one to emphasise the main ideas and results and put the connections between them very clearly. One keeps on doing this reviewing not only because the revision helps one to keep the old material in mind, but also because as one learns more one can understand relationships in other, deeper ways.

11.5 Ask questions

One of the aims is to teach the students to ask sensible question. If they are to do this, they would benefit by you asking suitable questions. For instance, if you have a theorem's statement you might ask the following obvious questions:

(1) Does this theorem seem reasonable?
(2) What does it mean geometrically? Algebraically?
(3) Can I supply an example?
(4) Why does it have such peculiar conditions?
(5) Which theorem is it related to?
(6) Could I generalise this theorem?
(7) What is the converse? Is the converse true?
(8) Where can I use this theorem in Physics, Chemistry, or Electrical Engineering (or whatever subject they are studying)?
(9) Is there an example which basically illustrates the whole theorem?

Of course you should not do this routinely. Doing it routinely would be tedious, and in any case would take too long. But from time to time you should study a theorem in this way.

Also from time to time you should get the students to ask the questions, by writing a theorem on the board and calling for questions. You will inevitably get some stupid questions. Do not be dismayed. When one is learning something for the first time it is natural that ideas are confused. What is required of you as teacher is to take the stupid questions and use them to help, because, with a little skill, even the stupid questions can contribute greatly to understanding.

11.6 Motivate

It is important to motivate your subject by relating to the needs and interests of your students. It is therefore much easier to do so if the students are relatively homogeneous. The juxtaposition of totally disparate students in a class is a serious blunder, and no teaching method can possibly overcome that disadvantage. The only solution is to ensure that there are different classes for these different groups. That is an expensive solution, but the other one, when a large proportion of students fail to learn much, is even more expensive.

For engineering students motivation is relatively easy. What is important for the examinations provides a strong motive, and so should be used frequently. For instance, "This theorem was used to answer question 4 of the examination in July 1996." Then one gives the solution. Something like that once or twice a week gives a further interest in the subject being studied.

Since the students are interested in engineering they need constant reassurance that the mathematics studied is going to be of value in their subject. It is helpful to have quotations from famous engineers who make this point. Or they could be less famous, but at least teachers on the faculty; some of those with a nice turn of phrase may allow themselves to be persuaded to make some sort of quotation. This can be shown as an overhead slide.

The director of undergraduate studies in the engineering subject concerned may be persuaded to come along when you give a quick survey of the course, and comment that this or that will be particularly useful.

A student who is studying a subsequent course, could also come along at one point and say, "Yes, that idea is just what I needed when I took course XYZ."

Once I was approached by two computer science students who had attended my linear algebra course. They asked me why I had not pointed out to them how useful matrices were for graphics, they being so delighted by their applications. I recorded a video of them and their enthusiastic comments.

Examples, say of functions involving x, could just as well for instance be examples involving I the current as variable, which gives a reminder that the subject is of value in electrical engineering.

Even better are examples which are of value in the development of the subject. Unfortunately these take a while to develop, and one can probably only have one or two in a course.

Another trick which takes less class-room time is to take a page in an important text-book in the engineering subject which uses the mathematical symbols or methods one is just explaining, and project it from an overhead transparency. For instance, partial derivatives. Or a point in that text-book where a differential equation is solved using one of the methods either just about to be discussed, or one that has been discussed.

A useful approach for adding interest, is to include historical asides. A photocopy of some historical aspectscan be sent round the class while they are solving problems on their own. Quotations are also useful (see Appendix C). The occasional cartoon adds a little humour (Appendix F). It may also be possible to find a suitable video. (See Appendix B.)

The use of these techniques should be sparing, otherwise instead of convincing the students it will simply sound like propaganda, and will take too much time.

It is useful to read, or at least glance at, some of the engineering books that your students will study. It gives one a good idea of the topics the student will need and also the way that engineers approach a subject. It will help ensure than important ideas and techniques that your students need will not be inadvertently left out. For instance, when I gave a course to aeronautical students, I looked at the book Anderson [1989], which I found extremely entertaining. I also found out that the curvilinear integral was an important tool in Anderson's book, but that this concept did not feature in the mathematics course.

For mathematical specialists motivation should be easier. It is worth while pointing out in advance the main results one is trying to prove, and why they are important, and some of the problems one can solve with the new ideas and techniques. Try to indicate how the subject studied fits in the whole picture. Include something of the history and the motivation of the mathematicians who originally discovered the subject. Videos, quotations and cartoons (Appendices B, C and F) are also helpful.

11.7 Introducing new concepts

If you introduce a new idea the most obvious and sensible question is "Why do we need that idea?" It is better if the student sees the need immediately rather than having to wait two lectures later. The authoritarian approach is to tell the student to shut up and wait patiently, since all will become clear later. I think that is a mistake. Try to give some answer, even it is not the best possible answer, so that what you are doing has meaning. Later on one can give a better reason, when it arises in a natural way.

Despite the above remarks, at times it makes sense to wait till one has learnt a little more about a topic, and this is also a strategy that the student must learn. This is especially so when a new subject is being introduced gradually, in small easily assimilable parts. At times it pays to have faith in the lecturer. Some topics are too complicated to be described simply in advance. But sooner or later, one needs to try and make sense of the topic in its entirety.

But if possible, I like to introduce ideas in a natural way. Consider for example the introduction of eigenvectors, certainly an important topic in linear algebra. Here are two approaches to introducing eigenvectors in R^n.

APPROACH 1: Eigenvectors.
Let A be an nxn matrix. A column vector $\mathbf{u} \neq \mathbf{0}$ is called an eigenvector with eigenvalue λ if $A\mathbf{u} = \lambda\mathbf{u}$, where λ is a real number.

The advantage of Approach 1 is that it is quick, accurate, and concise; it presents only the absolutely essential information. That eigenvectors will turn out to be a central part of the course becomes clear as the course continues. What they will be used for also becomes clear later on.

An alternative approach, which will be sketched now, has its advantages, namely it gives reasons for studying eigenvectors and relates the study of eigenvectors to the material that has been learnt earlier. Its disadvantage is that it takes longer to explain. Whether one should use Approach 1 or Approach 2 (or some other approach) depends on the situation and your judgement: such decisions are seldom clear-cut. But it is worth while considering the possibilities.

APPROACH 2: Eigenvectors.
In our last lecture we noted that a matrix could be regarded as a linear map L, and that generally speaking, L transforms lines to lines and parallelograms

to parallelograms. Thus in the following diagram the action of a linear map L on lines is indicated in Fig. 11.4, the line l being transformed by L to l'.

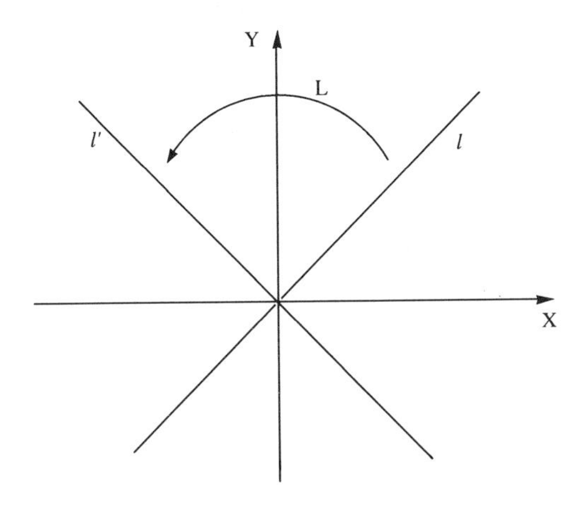

Fig. 11.4. L takes the line l to l'

The obvious question is, can it happen that L transfers a line back to itself, so that l' and l coincide? Since each line is determined by a direction vector, we can also ask what happens to such a direction vector **u**. It must be taken to another vector on the same line. But this implies that $L(\mathbf{u}) = \lambda\mathbf{u}$, where λ is a real number.

Thus we have the definition of an eigenvector: It is a non-zero vector **u** such that $L(\mathbf{u}) = \lambda\mathbf{u}$. If the linear map comes from a matrix A, we require $A\mathbf{u} = \lambda\mathbf{u}$. As an example, consider the matrix

$$\begin{pmatrix} 2 & 4 \\ 5 & 3 \end{pmatrix}$$

which has eigenvectors

$$\begin{pmatrix} 1 \\ -1 \end{pmatrix} \quad \textit{and} \quad \begin{pmatrix} 4 \\ 5 \end{pmatrix}$$

with eigenvalues -2 and 7 respectively.

Now the remarkable thing is that this definition of an eigenvector which comes from a simple geometric consideration, namely transforming a line back to itself, has a lot to do with the real world. In particular it is concerned with resonance.

Take a plucked string, for instance a violin string. Its eigenvalues correspond to its sound frequency. Suddenly an abstract mathematical term has something to do with music.

Resonance is an important concept. It is quite standard when considering a mechanical system to calculate eigenvalues so as to ensure to avoid these and thus prevent your system shaking itself to pieces.

You will find eigenvalues and eigenvectors appearing in the solving of systems of differential equations and you will find them in Quantum Mechanics. These concepts are thus of great importance.

So eigenvalues are of importance in the real world and we would like to know given a matrix A how to calculate them. In this we are fortunate to have a powerful ally in the form of the concept of determinant which we developed earlier in the course.

11.8 Informal and formal definitions

If at all possible give, to begin with, a "rough intuitive" definition, which you can explain is not the proper definition, but is in fact a useful concept to have around, both to get a feeling for the subject, and also to suggest how to go ahead proving something properly.

Thus I suggest (as I have before in §6.7) that one can profitably give two definitions of a new concept, the one an informal, preliminary one, which is so labelled. The other, the correct mathematical definition. One should also show that the correct mathematical definition is needed. A simple example is the rough definition of a continuous function as one whose graph can be drawn without lifting pen from paper. Ask the students to use this definition to prove that the product of two continuous functions is continuous. This way they will realise that the proper definition has a point.

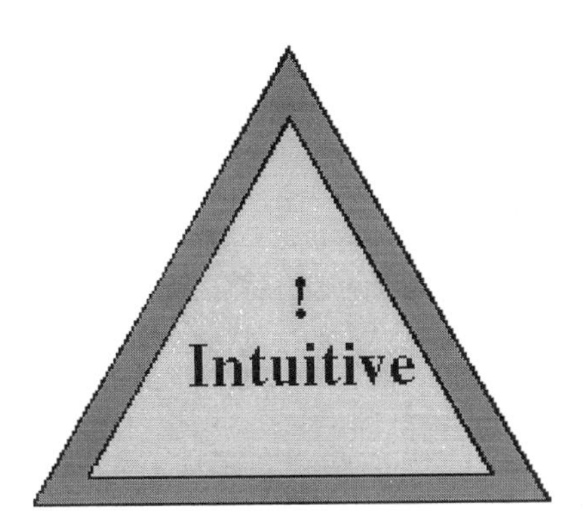

Fig. 11.5 Danger, an intuitive explanation

I like to mark the intuitive idea with the symbol shown in Fig. 11.5, which is supposed to symbolise the intuitive idea and also danger, the danger being that one takes the intuitive definition as the correct definition. One must emphasise that definitions marked with this triangle can not be used to give an acceptable argument.

When introducing a new term, try to give an explanation for the name, so that it is easy to remember.

11.9 The Good Picture and the Essential Example

Very often a **good picture** is an excellent way to explain something. A well-known example is the proof that **N**x**N** is equipotent with **N**, the natural numbers, as shown in Fig. 11.6.

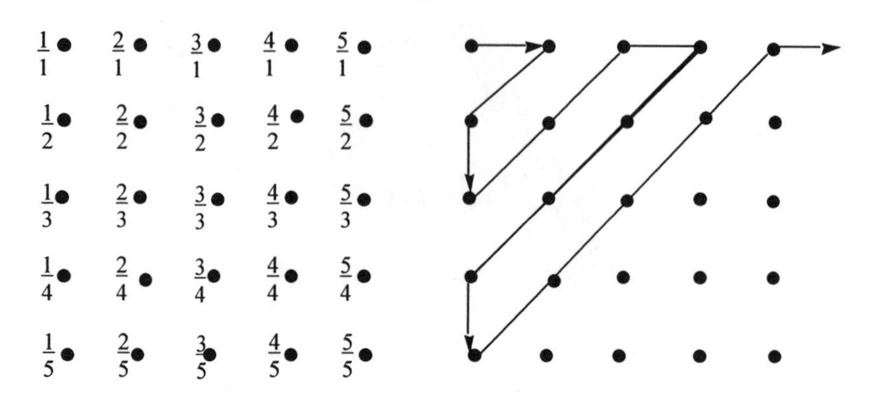

Fig. 11.6 **N**x**N** is equipotent with **N**

Another good way of proving **N**x**N** is equipotent with **N** is by simply defining the function $f(i,j) = 2^{i-1}(2j-1)$.

However, simply defining $g(i,j) = j + \frac{1}{2}(i+j)(i+j+1)$ and then checking that it is bijective is not a good way to prove equipotence of **N**x**N** and **N**.

The point I am making is that although these three proofs are all correct, the last one provides no understanding; it simply provides one with something to check.

The picture can provide most of the insight and conviction in a theorem. Littlewood gives this example, Fig. 11.7, in his Mathematician's Miscellany.

Here we have an increasing continuous function $f(x)$ defined on an interval [0,1] and having [0,1] as co-domain. We choose x_1 to be any number lying between 0 and 1, and define $x_{i+1} = f(x_i)$. Then x_i tends to a fixed point as i tends to ∞.

Very often a good example is the clue to understanding, indeed the example is so good, that it could even be called The Essential Example, for it provides the very essence of the understanding. An illustration of this is the one-step compactification theorem.

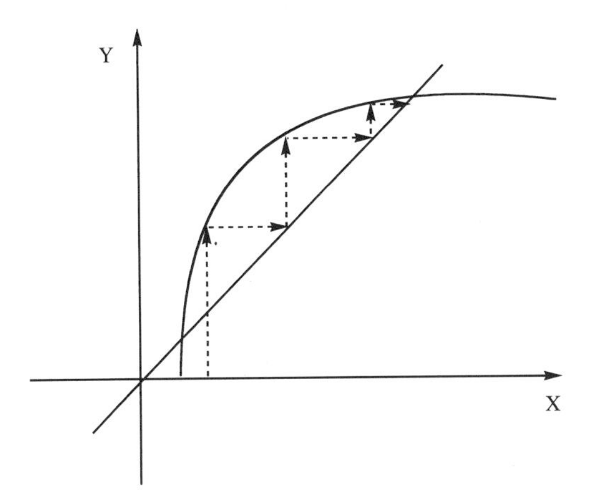

Fig. 11.7 The fixed point theorem

Given a non-compact topological space X, we can add a new point ∞ so that we form a new topological space X* which is compact. The illustrative example is shown in Fig. 11.8 where we have as X the real line, which is not compact. We then add a point P outside the line, and draw the circle S shown in the figure. The points Q on the circle are mapped onto X by drawing a line through Q from P to meet the real line in the point Q´. Clearly in this way all the points on the circle are mapped onto the real line, with the exception of P, which we map onto ∞.

We note that the open intervals (a,b) of S which do not contain P correspond to open intervals of the form (a´,b´) on X, while the open intervals (c,d) which contain P correspond to the complement of the closed interval [c´,d´]. Since the closed interval [c´,d´] is compact, we are led to the following argument in an arbitrary space X.

We add a point ∞ to form the new space X*. The open intervals of this space are the open intervals of X together with subsets of X* which contain ∞ and whose complement in X is compact.

We can proceed with the arguments in this new space X* constantly seeing its meaning and significance in the example. It makes the theorem much easier to follow and understand.

It is worthwhile keeping these methods in mind when trying to explain difficult theorems and ideas.

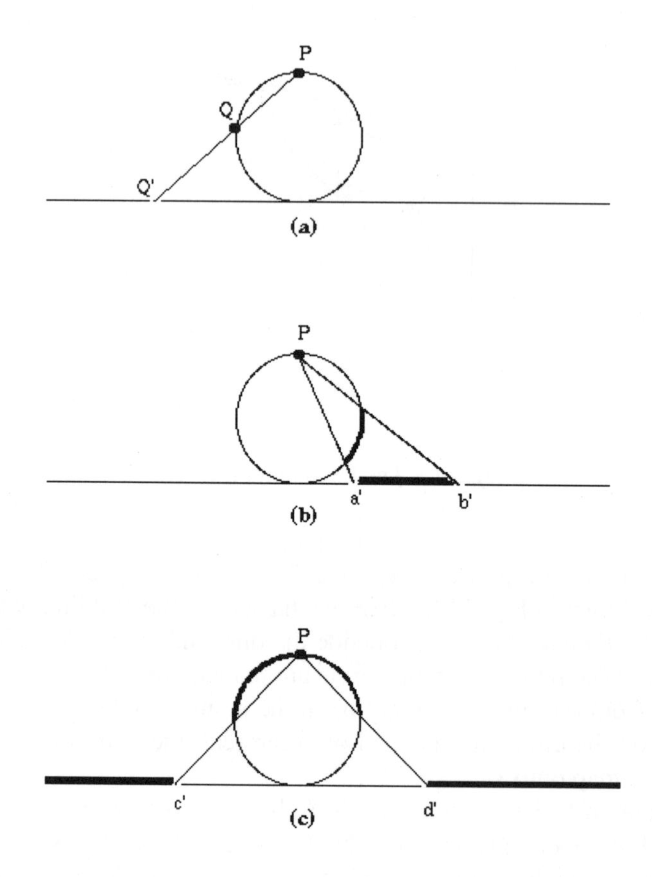

Fig. 11.8 One point compactification

11.10 Repetition and maxims

It is essential to remember the importance of repetition. It is better to say things twice rather than once, and thrice rather than twice. It is also worthwhile reminding students of concepts or definitions or theorems they may have seen previously and will be needing now. Avoid the trap of trying to say everything in the most economical way.

There are certain ideas, techniques and procedures that one would like to emphasise over and over again. Since repetition can become boring, it helps to make up some maxims or rhymes that can be repeated again and again. See Appendix G for examples.

11.11 Degree of generality

Some like to give the most general form of a theorem or result. This can be the clearest and best approach, but the most general form can be artificial, difficult to motivate and difficult to prove. An example of this occurred when one of my friends took over my course on Vector Spaces. He told me he was going to teach a slight generalisation, finitely generated modules over a principal ideal domain. He assured me that in Germany this sort of more general approach would be perfectly effective. However, in England, he was forced to admit ruefully later, it failed completely.

Kaplansky [1969] takes precisely the opposite approach to my friend. Kaplansky begins with Theorems 1 to 14 which are concerned with abelian groups. Then he makes the blanket assertion that all these theorems hold for modules over principal ideal rings. He points out that only nominal changes are required in order to prove these theorems, and he continues without giving detailed proofs of the results. I think this is an admirable approach, a saving of time and effort and yet a procedure which still leaves the reader with a good understanding.

In general I take a simple and easily stated version of a theorem, and if need be, give the more general result later.

I also find it useful to make an argument concrete by taking a particular case. For instance, if a matrix A has eigenvectors with distinct eigenvalues then the corresponding eigenvectors are linearly independent. I give the proof for the two eigenvalues 2 and 3. Despite the fact that this is only a minor simplification of the case of two eigenvalues λ_1, λ_2, students find it very helpful. Replacing λ_1, λ_2 with numbers means that there are two fewer items to worry about. This is an advantage partly because this makes the proof a little more concrete, and also because then I can say the same thing twice when I prove the general case of two arbitrary eigenvalues. Of course that then leaves the proof for n eigenvectors still to be proven by induction. How do I manage to do all this in the limited time available you may ask? Well, of course I do not. Sometimes I simply say that the general case may be proved in a similar way. It is always a compromise. It also depends on the students. Some can readily absorb the general case with proof by induction without any preliminaries.

In Griffiths [1974] it is pointed out that there are various styles of explaining ideas. For instance, one can say that d/dx is a linear operator, or that $d/dx\,(f(x) + g(x)) = f'(x) + g'(x)$ and that $d/dx\,(af(x)) = af'(x)$ or simply take a number of examples of functions $f(x)$ and $g(x)$ and calculate the derivative of the sum directly, allowing the student to tacitly infer the rule that way. Perhaps the reason why it is easier to understand the last approach is that one actually does something concrete. In the same way taking the particular eigenvalues 2 and 3 helps make the understanding more concrete.

11.12 Organising breaks in a lecture

A lecture period of an hour is quite common. But most students find it difficult to concentrate for the whole hour, so it is useful to organise a break in the middle of the lecture. A natural break is of course best, but any sort of break is good. I like to give a brief summary both before and after the break.

The breaks I use are the following:

- When I come to a routine calculation, e.g. some algebraic computation, factorising, or solving a quadratic equation, I stop lecturing and ask the students to do the calculations.
- I ask the students to take a break for a minute, to stretch and to yawn.
- I ask the students to have a five minute discussion with the students sitting beside them.
- I ask the students to briefly review the lecture so far, by jotting down the main points.
- I tell a joke. (Use this method only occasionally and make sure the joke is relevant to the mathematics.)
- I arrange a demonstration. For instance, when discussing conic sections, I bring a torch, darken the room, and shining at various angles, obtain ellipses, circles and hyperbolas. (One needs to experiment before-hand with different torches, not all work well.) When talking of permutations and combinations I ask four students to step up and explain the concepts by permuting the students.

- I ask the students to guess the result of some calculation or idea. (§12.8 describes this method.)
- I change my style of presentation, e.g. from straightforward lectures to presenting examples, or I change to questioning the students.

11.13 Tutorials

Tutorials are student directed and form an important part of any course in mathematics. The main feature of tutorials is that they involve only a small number of students and therefore are more personal. Thus your relationship to and the way you treat each student becomes of great importance. Students differ greatly. For some you can be casual, others prefer a more formal approach. Most students are insecure. They may be frightened that they have failed to do the problems, do not understand the course and may not be up to it. So some reassurance is needed. Do avoid the sort of comment one famous, eccentric and brilliant woman mathematician used. She would stare with incredulity at a student's work. "Why did you do that, you silly boy?" she would demand indignantly, in her clear, penetrating voice, and the luckless student would turn a bright red.

Personally I look very carefully at the method the student is using, trying to see whether it can be pushed through successfully. If I am rushed I will say that the approach may quite possibly work and may be as good as mine, and it is a pity we do not have the time to check. The thing is to treat what the students write with respect. Ridicule what they do and it will probably inhibit them in the future. But be truthful, and not patronising. Everybody understands when he or she is being talked down to.

Here just as in lectures you need to have pleasant personal qualities: respect, patience, the ability to listen and to pitch your conversation right to the student's needs, to encourage, to be generous, to admit your mistakes, and to be in good spirits. Mathematics should be fun, and it can't be if you are not cheerful. Here again we have come to one of those cases where some can do the exact opposite and get away with it. They can be rude, sarcastic, challenging, aggressive, and still get the best out of their students. How they do it, I do not know. For most of us I recommend trying to be pleasant instead.

Encouragement is required. There are plenty of useful things you can say. "I think you are coming along much better now." "Do you not find the things we discussed earlier in the term seem easier now?" "You have improved, I do not think that you could have understood that explanation when we began the course."

"I think that is a good answer." "That was a neater method than I had in mind." "Oh, I see you did it that way, I had not thought of that myself. Interesting, I wonder whether it will work. It's always worth trying." "No, it looks as if this method is going to fail. Pity, it seemed to be a good approach. Well one can never tell unless one tries."

Even small things can help. Just expressing a slight interest in the student helps. "How are you getting along?" "Did you find the last lecture intelligible?"

11.14 Various types of tutorial

Tutorials differ widely, and perhaps it is helpful to at least indicate some of these various types. There is for example prep, problems classes, discussion tutorials, individual consultations etc.

In prep students sit at their desks and work through problems, while the lecturer circulates, and provides help, usually in the form of hints as to what the student should try if stumped. This form of tutorial is easy to handle, since the lecturer simply responds to the student's questions. As indicated above, it is better to consider the student's method and point of view, rather than straight away give a faultless and elegant solution. If and when one does give a brilliant solution, it is worthwhile saying something like "I am rather proud of that solution. It took me the whole of yesterday to find it." I say such things because students assume that they too should be able to write the answer out neatly and in faultless form in five minutes, just as their tutor does, without taking into account that the tutor may very well have done a large number of such problems in the past, or worked the problem the day before.

If the group is large, one may not be able to answer all the students' questions in the time available. I have found two strategies useful: if I have just given help on a problem and another student asks the same problem, then I refer the one student to the other. My other strategy is to bring two loose-leaf files into the class, the one has hints to the solutions of the various problems, the other has complete answers. I then can refer the student to these files when I find that I am overwhelmed with questions.

There is also the difficulty of slips in calculation. Often students want very much to know why they have not got the correct answer. It is worthwhile trying to find out where they went wrong, but if one is short of time, I simply check their method, and if correct, suggest they leave the problem and return to it the next day when they will probably find the small slip they have made. I tell them if the problem still is not resolved to come back to me.

Then there is the problems class. One method is simply to take up problems the students have had difficulty with and solve them on the black/white-board. If it is the lecturer who simply gives model answers, then it is in my opinion important for the students to have tried these problems before hand. What I generally do is call for problems that the students want solved, and then a day before the problems class, give a list of those problems I will tackle, and ask the students to try them before coming to the class.

Of course if the class is small, the tutorial can be conducted in a more flexible way. The whole class can discuss how to solve the problem and one can get to discuss various methods and strategies, the students can provide the answers and so on. One can use a sort of Socratic discussion, following the excellent examples given by Polya [1965].

One lecturer told me how he once spent an hour discussing how to obtain a sensible definition of the dual of graph. One student criticised that tutorial, saying that a perfectly good definition could have been given in five minutes instead of wasting time. But the idea of discussing definitions in this way seems very good to me; definitions are not God given, and it is important to realise that the right definition is hard to come by and is of great significance. Perhaps the student had only the one aim of learning sufficient material to pass the course, and then of course his comment would be valid from his view point.

There is a lot to be said for class discussions. It depends as usual on the students. Some are impatient, they do not want these discussions. They want really to be told the solutions. The sort of student who is deeply interested in mathematics will appreciate the discussions and learn a great deal from them. Others who simply want to know how to solve standard differential equations for use in their engineering courses will not be pleased.

It is useful to set aside some time when the students can come to consult you. Check whether the suggested time makes it possible for them to come.

Also agree to them making appointments at other times when the set time is not suitable.

11.15 A price to pay

It is important to bear in mind that all suggestions for improving teaching have disadvantages as well as advantages. It is inevitable that for instance giving a good introduction and good summaries will require time that must be taken from other desirable aims, and thus you may find that in consequence you must leave out the details of a proof or skip over a good example. The art of the lecturer is to find the right balance.

11.16 Summary

This chapter serves to emphasise many of the ideas already discussed in the previous chapters, in particular the rules of teaching in Chapter Seven.

The lecturer must be on the student's side, and must make the student feel this too. How the results came about, how one can remember them, why they are important, both in the development of the subject and for the students' own interests, the need for questions, and the importance of overviews and summaries, are all important. Nevertheless time is short, and the lecturer will have to make compromises and will find it difficult to include all that is desirable.

12. Some Practical Points

Preview

This chapter is meant to provide the individual lecturer with some useful general teaching techniques, such as how to discuss the statement of a theorem, the use of juxtaposition, notation, and encouraging the students to guess and ask questions. It is important to be aware of sections of the course that students usually find difficult, for these sections will need extra preparation and ideas.

12.1 Your voice

Your voice is an important part of your teaching. It must project. If it does not, practice humming, in such a way that the lips vibrate. You may not have a beautiful voice, but you should have a voice that carries clearly. If you have time, I suggest you look at books on the theatre, or how to give speeches, which contain exercises for the voice.

You must also learn to vary your voice. The worst thing you can do is to speak in a monotonous, unvarying pattern. You must vary between speaking slowly and a little faster, you must place emphasis on certain words. You can speak louder, or more quietly. You can change your tone. But whatever you do, you must vary.

If you are interested in the material, you are likely to show that interest.

Then too you must have a strong command of the English language. This can be achieved only by widespread reading, and attending literature and writing courses. I fear this suggestion is one that is impractical, since few people would have the time or interest. Nevertheless, it is a fact that many mathematicians do not have an extensive vocabulary, nor a fine appreciation of language, and without this it is difficult to make your instruction interesting.[11]

Take a look at the writing of Richard Dawkins in for example Dawkins [1996]. In this book he manages to use language well. The standard techniques of writing appear here: metaphors, striking phrases, imaginative analogies, sentences with interest and tension, which urge the reader onwards. Another good example is that of Jeans. How for instance does he explain the expanding universe? He blows up a balloon. There must be many ways in which we as mathematicians can express ourselves better.

In Allenby [1983], Dunham [1994], Gullberg [1997], Stillwell [1989, 1992] we have excellent illustrations of how one can write mathematics as an account that is at once interesting and mathematically informative. It can be done, but it does require the skills of a story teller as well as that of a mathematician.

12.2 Notes

Neat handwriting on the blackboard/white-board is a distinct advantage. Spend a little time practising. Like anything else one can improve. Write more slowly. Check by going to the back of the class whether what you write is easily visible from all parts of the hall.

It's worthwhile having a consistent set of headings and subheadings, so that these alone constitute a summary of the whole course. So you should consider how these will read on their own.

Your notes should be written in note form, not in flowing sentences. So you should space out clearly the conditions from the conclusions and instead of listing items one after the other with an "and" between them, number them as separate points.

Give each theorem a brief name or description. Then you can refer to the theorems both by number (Ch1, Theorem 4), and by the short phrase (f.g. abelian structure theorem).

It is also helpful to give tables, and inter-relationship diagrams.

Each result should have some geometrical representation if at all possible.

[11] This paragraph had to be corrected in three places, thus unintentionally emphasising my point.

Use lines to point out important points. E.g. in Fig. 12.1 I have emphasised that the interval must be closed and that the result is not necessarily true if the interval is open.

If $f(x)$ is continous on the closed interval [a,b]

N.B. must be the closed interval

Fig. 12.1 Pointing out important points

In general one can make comments very easily in this way. Thus in Fig. 12.2 it is possible to emphasise how partial integration works with the additional lines indicated; one can furthermore use colour to emphasise these points even more strongly.

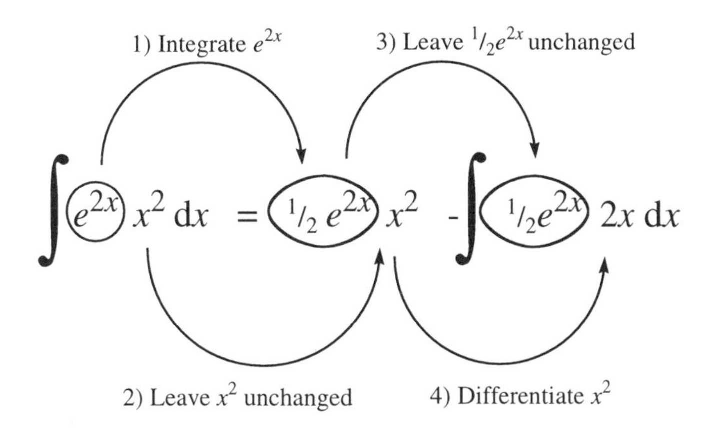

1) Integrate e^{2x}
3) Leave $^1/_2 e^{2x}$ unchanged
$\int e^{2x} x^2 \, dx = {}^1/_2 e^{2x} x^2 - \int {}^1/_2 e^{2x} \, 2x \, dx$
2) Leave x^2 unchanged
4) Differentiate x^2

Fig. 12.2 Emphasis with extra lines

A decent sized blackboard will enable you to write two items to be contrasted side-by-side. Thus, if I want to introduce Taylor's theorem for two variables, I draw a line in the middle of the blackboard, and write

Taylor's theorem for one variable on the right-hand side of the board and the theorem for two variables on the left-hand side of the board.

A proof of a theorem can often be made much clearer by starting on one side of the blackboard and writing a proof for a particular case, in other words, an example. Then adjacent to that write out the formal proof, using exactly the same steps and techniques for both. This method of juxtaposing the familiar with the less familiar is very useful, and shall be mentioned in the next section.

12.3 How to discuss the statement of a theorem

There are a number of typical ways of discussing the statement of a theorem, among which I include the standard method, the whole story method, the detective story method, and the barrage of questions method.

I begin with the standard method. Draw a line down the blackboard and on the left give an example to illustrate the theorem (i.e. use juxtaposition as mentioned in the previous section). Present the example in a form which enables it to be readily compared with the statement of the theorem. It is best if you have a very simple example, which you always repeat when you describe the theorem. (For instance, $f'(x)$ increasing in an interval implies $f(x)$ is convex. I always couple with the example of the parabola $y = x^2$, not only when I introduce the theorem, but whenever I refer to or use the theorem.)

- On the right of the line give the statement of the theorem.
- By means of captions point out the conditions and conclusions.
- Comment on any peculiarities of the conditions and conclusions.
- Try to give a short descriptive title to the theorem.
- Indicate the geometrical and analytic significance of the theorem.
- Give another example. Relate the theorem to other theorems and ideas.
- Explain the intuitive content of the theorem. Explain why it is reasonable that the theorem is true.
- Say what the theorem can be used for.
- Give a demonstration that the theorem is true. Only then should you give a proper proof.

The next approach is one mentioned by Krantz [1999] which he attributes to Halmos.

Krantz explains that if you want to talk about the fundamental theorem of algebra, namely that all polynomials have roots over the complex numbers, you could on the one hand simply state the theorem, and then prove it, which he feels is inadequate.

Instead you should tell the whole story. Explain that linear equations can easily be solved. Explain that quadratics have a solution. Deal also with cubic and also quartic equations. Then explain that there is no similar formula for solving quintic equations. Finally state the fundamental theorem.

The "detective story" method was used to great effect by Reinhold Baer. He would set the stage, pose a problem, and proceed to solve it. This clue would arise, and then another, suddenly difficulties would arise and the problem would become even more intractable. It was expressed very much like a detective story, and was most effective.

When talking about the least natural number for which a proposition was not true, Baer used to talk about "the least criminal." For instance, suppose one wants to prove the existence of a root (real or complex), to a given polynomial. One can put it this way. Suppose there are such polynomials without roots, and that $f(x)$ is such a nasty polynomial which does not have a root, i.e. $f(x)$ is never 0. Of course $f(x)$ is a criminal. Well then, if at least one inverts $f(x)$ one should get something quite pleasant, and indeed one does, $g(x) = 1/ f(x)$ is well defined in the region, since $f(x)$ is not 0. In other words, $g(x)$ is a fine citizen, a model of rectitude, especially as it is analytic. This means we can consider etc. etc.

Here is another demonstration of this approach, in discussing diagonalisation of symmetric matrices. We begin with an example of a 3x3 symmetric matrix A which turns out to have three distinct eigenvalues. And we find a matrix P using the corresponding eigenvectors. The next 3x3 symmetric matrix B taken as an example has two distinct eigenvalues, 2 and 3. The discussion proceeds as follows:

The method for the matrix A worked extremely well. We try to do the same for the matrix B. The corresponding matrix Q of eigenvectors does not work, because $Q^T \neq Q^{-1}$. This is because the eigenvectors are not orthogonal. We have two eigenvectors with eigenvalue 2 and one with eigenvalue 3. We note that the eigenvector with eigenvalue 3 is orthogonal to the others. But the other two are not orthogonal to one another. We seem to have come to a dead-end. Will we have to give up in despair?

Perhaps we can try again. To summarise, we have two eigenvectors with the same eigenvalue. Can we get two orthogonal eigenvectors from them? Perhaps we can modify the second eigenvector by adding a multiple a of the first to it. Can we choose a so that the vectors are now orthogonal? Let us try it, then [after some calculations]...this seems to work. We now have two orthogonal eigenvectors and we can go back to trying our old method. We normalise the eigenvectors and take the corresponding matrix Q which we see is orthogonal, and it does in fact diagonalise the matrix B. It seems to work perfectly. We need to see whether we can use this method in general.

The final method I want to mention I learnt from my brother Gilbert Baumslag. I call it the "barrage of questions method." After introducing some concepts, my brother would propose a number of questions. After he had raised them, he would gradually proceed to answer them.

Here is an example (from linear algebra, and inspired by Kemeny [1964]):

Thus (to summarise) we have now visualised a single equation in three unknowns as a plane in three dimensional space. If we have two such planes they could be parallel, or else they could intersect in a line. And if we had three such planes, all three could intersect in a single point, and we would have a unique solution to the equations. Or the three planes could intersect in a line, in which case we would have an infinite number of solutions, all expressed in terms of one parameter, since lines can be expressed in terms of one parameter. Or else they could intersect in a plane, in which case we would have an infinite number of solutions, expressed in terms of two parameters, since planes are expressed in terms of two parameters. Or else they could fail to intersect at all, in which case we would have no solutions.

Thus we have the following possibilities: no solutions, a unique solution, or an infinite number of solutions in terms of one or two parameters.

So we ask ourselves the same sorts of questions for n equations in n unknowns. Can we in fact get the concept of n-space just like we have 3-space, and will n equations in n unknowns have a variety of possible solutions, sometimes a unique solution, and sometimes no solution, and sometimes an infinite number of solutions expressed in terms of s parameters, where s lies between 1 and n—1? And how would we know, given a system of equations, just which of these would be obtained?

12.4 Notation

Choose your notation with care and make it memorable. Point out ways of remembering the notation and don't forget to mention the obvious, e.g. *there exists* is a backward E, Z stands for the German zahl which means integer, etc.

If I am discussing a matrix I will if possible refer to an r x c matrix rather than an m x n matrix, simply because r reminds me of rows and c of columns (m & n can be confused).

Similarly Littlewood suggested to Hardy to use L, R (standing for left and right) for a section in defining the real numbers. Hardy thought the suggestion not worthy of acknowledgement, but Littlewood thought (and rightly) that it was a significant improvement.

A symbol I have found quite useful is the use of a dot • on the left of a symbol to denote a set. Thus if we use the convention that mere elements are indicated by lower case letter such as a, b, c, etc., then •b is a set whose elements consist of things like b, i.e. ordinary elements which themselves are not sets, whereas **:**b denotes a collection of sets. The point is that if one removes one dot from **:**b, one gets a typical element, i.e. •b, itself a set of elements, a typical one being b (obtained after removing a dot from •b). If we use the convention that upper case letters denote sets, e.g. A, then •A will denote a set which consists of elements which themselves are sets. **:**A denotes a set which has as elements sets of sets. Thus I would denote a topology for a given space X by •O, i.e. the collection of open sets, which is a set consisting of sets.

Mathematicians tend to be conservative, and use the notation and methods which they were taught. Often it is clear that a poor method has remained simply because it appeared historically in that way. Be prepared to change notation and techniques when they are obviously not very good.

You may object that if you take so much time and effort as I have suggested above, with careful and lengthy explanations, summaries, and examples, that you are very unlikely to finish the syllabus, and that is true. You will find that you will have to be very selective in the methods you use.

12.5 Equivalence relations and functions

Some topics one knows from previous experience will prove to be hard going. It is best to think carefully about these well in advance. The quality of your course will be *immeasurably improved if you can identify the tricky topics* and find ways of handling them. Here are some examples:

Equivalence relations

Equivalence relations are an example of something which appears to be trivial but at the same time perplexingly abstract. Students can be totally confused because we don't seem to do anything with the concept. So either don't bother with the idea or else use the concept of equivalence relation for something.

I think an excellent example of equivalence relation is the relation C: aCb if a and b have the same colour. The three conditions of an equivalence relation then are natural enough and lead to the definition. The equivalence class containing a consists of all objects with the same colour as a. Thus equivalence class appears in a natural way.

One can use the equivalence relation congruence modulo p to define the fields of order p, at any rate say for p = 3 or p = 5.

One can also define the complex numbers as equivalence classes of polynomials in x, where two polynomials are equivalent if their difference is divisible by $x^2 + 1$. This is a very striking example, which is worth using even if (which often happens) the students do not really understand it. It does indicate that the subject is worth studying.

One can also use equivalence classes for counting arguments.

Functions

Functions are also tricky things to define. It is very tempting to define them as subsets of a cartesian product and to get the whole business over promptly. It does not normally work very well that way. So I prefer to introduce functions gradually, restricting discussion to polynomials, rational functions, trigonometric functions, $\sin(1/x)$, $\sin(x)/x$ and step-functions.[12] Thus it is possible to bring out the importance of domain and co-domain.

Inverse functions I introduce by examples of specific functions. I also draw a graph and show how the inverse is obtained by the reverse process used to find the value of the function. Thus $f(x)$ is obtained from the graph

---

[12] I strongly urge the students to remember this list of functions in accordance with §6.7.

by drawing a line parallel to the y-axis upwards till it meets the graph, and then going parallel to the x-axis till it meets the y-axis, I put arrows on this diagram, to indicate the way we go. The inverse is then obtained going from the y-axis in the same way, but in the opposite direction, that is, a simple reversal of the process (Fig. 12.3).

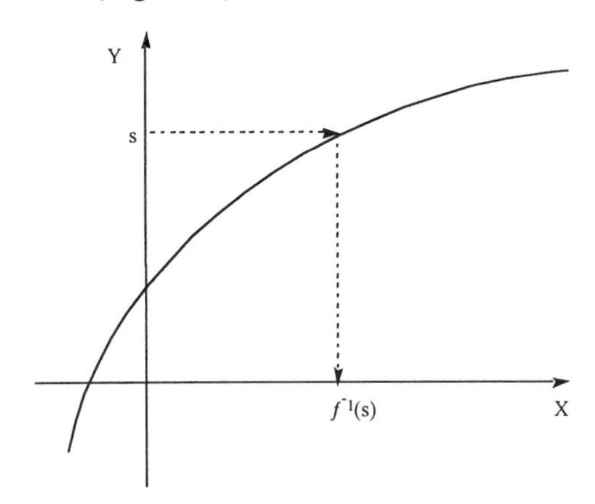

Fig. 12.3 The inverse function

Next I discuss whether this geometrical method always works. Thus I consider examples of functions which do not have inverses, which leads in a natural way to injective and surjective functions. Thus, on the graph, injective means that every line parallel to the x-axis meets the graph in at most one point. Surjective means that every line, parallel to the x-axis and passing through a point on the y-axis that belongs to the co-domain of the function, meets the curve at least once. If one uses Venn diagrams to represent functions with arrows indicating the action of the function, then an injective map has for each point in the co-domain at most one arrow, and a surjective map has for each point in the co-domain at least one arrow.

Of course the alternative method is to begin with the definitions of injective and surjective, and then to show that these are necessary and sufficient conditions for the existence of an inverse. This is the old method, which I would avoid in a beginner's course, but I think it is most helpful to have it at some stage.

One of the reasons why the old method is not always successful, is that the students fail to see the point of the new definitions. Why are they there?

They seem just to be mixture of words. Things were even worse when we used to speak of onto and into functions, which were very confusing.

12.6 Limits

If there is one concept that almost always fails to be understood by the average or below average student, it is that of limit. There is no concept more difficult to explain. How should it be done? I think the greatest difficulty is to try to imagine some sort of context to make the definition meaningful. I have three basic approaches:

The first method

The first method is concerned with a limit as n tends to ∞ motivated through approximation. For this I introduce the method of producing an estimate to the square root to a given number. So if one wants to find $\sqrt{a}$, we take a first guess x_0, and then thereafter define

$x_{i+1} = (x_i + a/x_i)/2$. The questions is how far one needs to go to get a good estimate. This of course depends on how accurate the result must be. If the estimate must be accurate to 10^{-3}, then the number of calculations one needs is a function of 10^{-3}, say $N(10^{-3})$ Obviously we will need a larger number of calculations if the maximum error is 10^{-5}, this is say $N(10^{-5})$. In general we have a good method of approximating $\sqrt{a}$ if for every suggested degree of accuracy ε, we can find an $N(\varepsilon)$ so that x_n is within the error ε for all n larger than $N(\varepsilon)$. In this way we get an interpretation of the definition of limit as n tends to ∞. Armed with a scenario which the student can help reinforce intuition, it is easier to understand the definition.

The second method

Next we come to discuss the limit of $f(x)$ as x tends to a from below. I describe it informally as follows:

Intuitive definition (the "expected value definition"): The limit of $f(x)$ as x tends to a from below is what you would expect $f(a)$ to be from your knowledge of what $f(x)$ is for all $x < a$.

Note that even if one knows the values of $f(x)$ for $x = a$ or $x > a$ one disregards them in defining the limit from below.

In reality this is no more complicated (nor easier) than betting on the horses. One knows how each horse is performing before the start of the race, say time t_0. With the knowledge of these performances one can work out who the winner ought to be. Unfortunately even with the best care in the world, we are not able to work out what will actually happen in the race. That is, the limit as t tends to t_0 may not equal the actual value of the function at t_0. Although this analogy can not be pushed too far, it is helpful, and it is worth remembering when the concept of continuity is discussed.

To get a feeling for what expected value means, I look at various functions. $F(x) = \sin(1/x)$ is an obvious example. As x tends to 0 from below we see clearly that the function oscillates, and that there is no expected value, so it has no limit. $G(x) = \sin(x)/x$ is also a good one to look at. I use a spread sheet program to work out values of $\sin(x)/x$ as x gets closer and closer to 0, and it soon becomes apparent that the limit looks as if it would be 1. Finally, I consider the step-function $H(x) = 1$ if $x \geq 0$, $H(x) = 0$ otherwise, and the limits as x tends to 0 from below and above.

The third method

The third method is considering the building of a tower or manufacturing something.

If we want to build a tower with bricks of size a it is impossible in practice to get a brick of size exactly a, one can in real life only get bricks within a certain degree of accuracy.

But we can ensure that the bricks are of size a within an error margin of δ, i.e. at most $a + \delta$ and at least $a—\delta$. If we have the tower is built of so many bricks, say 1000 high, and we want the final building to be a height of $1000a$ within an error margin ε, then we will have to choose the error margin of the size of the bricks to be within $\delta < \varepsilon/1000$. In other words, we decide what margin of error we want in the building, and then we choose the error margin in the size of the brick to be δ where in this case

$\delta < \varepsilon/1000$.

Similarly, suppose we have another manufacturing process in which we begin with blocks of size x and the final result is an article of size x^2, and we would like an article of size a^2, with say $a = 3$, i.e. we want an article with size 9. We would like $a = 3$, but we decide on the error margin in the final result, of say ε. What must we choose for the error margin δ in a to ensure that we are within this error margin?

If we choose $\delta = \varepsilon/7$ we will see that the error in the final product is at most ε. (When explaining this to the students, the calculations, naturally, are done to indicate that this choice of δ will do the trick.)

Now the above translate rather easily into the definition of limit as x tends to a. Incidentally, although it is a minor matter, I like to have used the Greek letters ε and δ earlier in the course, so that the student is used to writing them and pronouncing them long before we consider limits.

Turning now to the definition of continuity, it is worthwhile discussing some discontinuous functions first. I like also to recall the horse race analogy. Recall that we knew the results before the race, but the tendencies do not necessarily result in the predicted winner. Thus I have my

Intuitive definition of continuity ("no surprises definition"):

f(x) is continuous at a point a if we get no surprises at $x = a$, that is to say, f(a) is defined and is what you would expect it to be by studying f(x) both before and after the point a. More explicitly,

$$\lim_{x \to a-} f(x) = f(a) = \lim_{x \to a+} f(x).$$

12.7 Vector spaces etc.

An easy hurdle to avoid is giving similar sounding names for opposites which can be confused. For instance, convex and concave functions. When I introduce these concepts, I refer only to convex, and say that one should think of the example $y = x^2$. I then refer to the opposite without using the word concave. This avoids the confusion which arises merely because a student confuses the words. Choose one word or the other, and do not mention the other except as the opposite of the chosen word, at least for a little while at any rate. If each definition comes with its permanent example as discussed above, it is also easier to remember.

Linear independence and dependence are also definitions which prove difficult for beginning students to master. The first thing to avoid as already mentioned is that it can be easy to muddle up which one is independent and which one is dependent. Consequently I begin with the one, linearly independent, and stick to it for quite a while before introducing linearly dependent.

Definition of linear independence: Vectors $\mathbf{v}_1$,$\mathbf{v}_2$, ... $\mathbf{v}_n$ are said to be linearly independent if

$$a_1\mathbf{v}_1 + a_2\mathbf{v}_{2+} \ldots + a_n\mathbf{v}_n = b_1\mathbf{v}_1 + b_2\mathbf{v}_{2+} \ldots + b_n\mathbf{v}_n$$

implies $a_1 = b_1$, $a_2 = b_2$, …, $a_n = b_n$.

This seems more natural to most students, rather than considering the standard defintion that

$$a_1\mathbf{v}_1 + a_2\mathbf{v}_{2+} \ldots + a_n\mathbf{v}_n = \mathbf{0}$$

implies the coefficients are all 0.

As an example (not entirely accurate) I give the seasoning with three spices of a given dish. The taste of a dish seasoned in two different ways, one with x_1 gms of basil **b**, y_1 gms of salt **s**, and z_1 gms of pepper **p**, and the other with x_2 gms of basil **b**, y_2 gms of salt **s**, and z_2 gms of pepper **p** will be the same, i.e.

$$x_1\mathbf{b} + y_1\mathbf{s} + z_1\mathbf{p} = x_2\mathbf{b} + y_2\mathbf{s} + z_2\mathbf{p}$$

if and only if $x_1 = x_2$, $y_1 = y_2$ and $z_1 = z_2$.

In practice proving that we have unique coefficients is cumbersome, but the saving in confusion is well worth the awkwardness. One can in fact quite shortly go over to the standard definition, which in fact is an immediate consequence of the initial definition.

Thus observe that the definition of linear independence implies that, if

$$a_1\mathbf{v}_1 + a_2\mathbf{v}_{2+} \ldots + a_n\mathbf{v}_n = \mathbf{0}, \text{ since } \mathbf{0} = 0\mathbf{v}_1 + 0\mathbf{v}_{2+} \ldots + 0\mathbf{v}_n,$$

we must have that $a_1=0$, $a_2=0$,… $a_n=0$.

Another difficulty occurs when one defines an abstract vector space. Part of the problem is that there are so many laws that must be fulfilled. So I usually begin by concentrating only on two. Firstly, that there should be a sum, and the sum of two vectors must belong to the given set. And secondly, there must be a way of multiplying a vector with a real number, and the product too must belong to the set. Now these two are not sufficient to guarantee a vector space, just as an object with a steering wheel and four other wheels may not be a motor car. But if these two properties are not satisfied there is no possibility that we are dealing with a vector space.

Subsequently I add the other properties that are necessary. I do this with examples, for instance an example in which quite a few of the properties are satisfied but not for instance 1.**a** = **a**. This is because writing all the properties of a vector space at once intimidates the students.

Simple tricks work too.

I find it tricky to make 3-dimensional drawings. A couple of boxes, and balloons and perhaps other models help. I persevere with careful drawings of level curves and the projections on the three axes This is one case where I sometimes resort to the overhead projector. Maple produces very effective drawings, but if you want to improve your drawing skills, a useful book is Nelms [1966]. It is also a good idea to look at engineering drawing books, e.g. Earle [1989].

Three dimensional objects can be more easily visualised if one uses simple models. A stand with a rod placed on the floor can represent the z-axis, and one can draw the x- and y-axes on a large sheet of paper which is placed on the floor. Constructional toys can often be used to build suitable models.

To illustrate how one can calculate the volume of a solid, bake some strangely shaped bread, and then show how one can cut it into slices, and then estimate the volume by taking the sum of the volumes of the slices.

To get an idea of a surface or a solid body, one can also use level curves. Form these level curves with long strings of plasticine, and put the one on top of the other, to get a picture of the three dimensional body. These figures are too small for showing simultaneously to the class, but they can be handed around, so that the class sees how they can get a feeling for these 3-dimensional objects.

I have yet another simple method to represent these level curves, cutting cardboard shapes which I then position on an office spike. See Fig. 12.4.

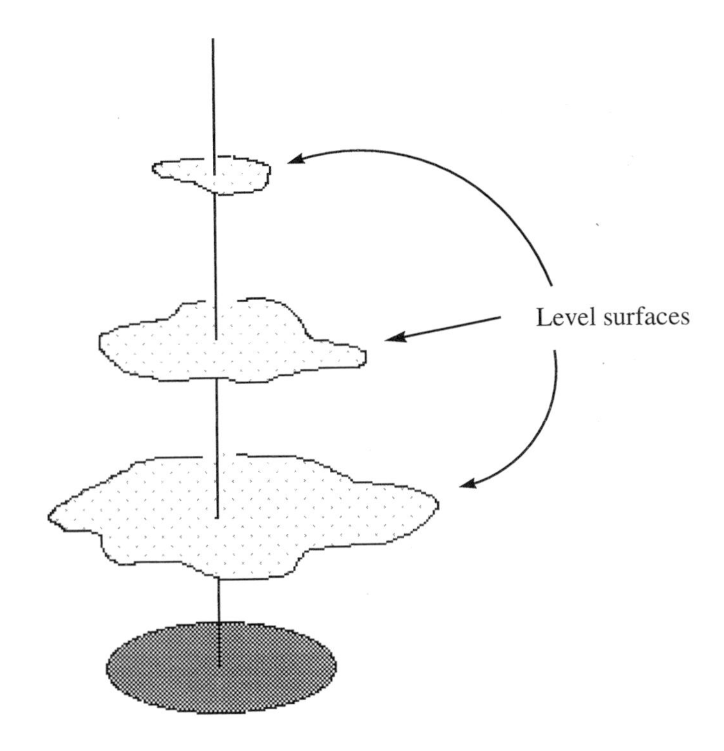

Fig. 12.4 Level curves on an office spike.

Sometimes a small trick can help a lot. For instance, in applying Euclid's algorithm (Fig. 12.5) I put a circle round each quotient. At the next stage one divides the larger of the two un-circled numbers by the other. This approach is also useful for back-substitution, when students often get confused as to what they are to substitute for. It is always the smaller of the un-circled quantities. Also by numbering these equations as one uses them, it is easy to avoid using the same equation twice, and also easy to check.

$$173 = \textcircled{3\cdot}\ 48 + 29 \qquad (1)$$

$$48 = \textcircled{1\cdot}\ 29 + 19 \qquad (2)$$

$$29 = \textcircled{1\cdot}\ 19 + 10 \qquad (3)$$

$$19 = \textcircled{1\cdot}\ 10 + 9 \qquad (4)$$

$$10 = \textcircled{1\cdot}\ 9 + 1 \qquad (5)$$

$$\text{Eq.(5)} \Longrightarrow \quad 1 = 10 - 1 \bullet 9$$

$$\text{Eq.(4)} \Longrightarrow \quad 1 = 10 - 1\,(19 - 1 \bullet 10) = 2 \bullet 10 - 1 \bullet 19$$

$$\text{Eq.(3)} \Longrightarrow \quad 1 = 2\,(29 - 1 \bullet 19) - 1 \bullet 19 = -3 \bullet 19 + 2 \bullet 29$$

$$\text{Eq.(2)} \Longrightarrow \quad 1 = -3\,(48 - 1 \bullet 29) + 2 \bullet 29 = 5 \bullet 29 - 3 \bullet 48$$

$$\text{Eq.(1)} \Longrightarrow \quad 1 = 5(173 - 3 \bullet 48) - 3 \bullet 48 = -18 \bullet 48 + 5 \bullet 173$$

Fig. 12.5 Euclid's algorithm, HCF(173,48)

12.8 Guess-check

I have already mentioned in §6.4 the method of guessing, which I encapsulate in the maxim "Do the mathematical two-step, guess-check, guess-check." It is important to emphasise the need for the check, that a wild guess on its own is unlikely to be of value, but a wild guess followed by a check to see whether the guess is valid or not, is almost always of value. For instance, in response to my request for a suggestion of a third vector **u** so that (1,2,3), (1,1,1), and **u** form a basis for R^3, one student hazarded **0**. It was easy to use this guess to show that adding **0** helps neither to get a linearly independent set or to find a set that spans the whole of R^3, thus enabling me to explain these fundamental concepts once again. Encouraging the students to take a guess is a technique strongly recommended by Polya, and I find it helps.

It also helps to encourage the students to ask questions. A stupid question is often a useful guide; it shows that I have completely failed to communicate to the students, or else it just gives an excuse for slowing down. As a variation, instead of asking for a question, I ask for a "dumb question." This can break the ice.

12.9 The teacher is wrong

At times you as the teacher will be wrong. Do not be ashamed of it. Admit it freely. Explain in detail why you are wrong. Explain how if you had been more sensible and thought more carefully about the question you would not have been wrong, or would have noticed earlier that you were wrong. Explain that it is human to make mistakes, but it is sensible to learn from your mistakes.

If you are so perfect that you never make mistakes, then I suggest you deliberately introduce at least one mistake per course. If you make more than a couple of mistakes per course, then you must prepare a lot more carefully. One or two errors are fine, but too many errors will lead to students losing trust in you.

12.10 Summary

In order to improve teaching one needs to consider all the matters raised in the preceding chapters. Teaching can be effective only if the whole department acts together, carefully designing time-tables and a choice of courses so that the basic rules of teaching can be fulfilled. Furthermore, the department should decide how much directed teaching should be given, and the students should be helped to acquire learning skills. The aim is that students should get practice in independent learning, i.e. acquire the ability to study on their own.

Some courses, perhaps most, should show the construction lines, and not only the neat, sanitised, and finalised versions. Others (preferably the more advanced courses) can be formal, concise and given in the axiom, definition, theorem style. In general, a variety of teaching techniques (group teaching, seminars, self-study etc.) and a variety of teachers will be an advantage. However, lecturing itself remains an important method of teaching, and teacher-centred teaching has many advantages.

If the courses have been well thought out and organised by the department as a whole, the individual lecturer can then do his or her part. If the system has not been well thought out, there is very little that the individual can do.

The tricks and tips in Chapters 11 and 12 are naturally merely examples, which will not work for all classes and students and in any case may require too much time. Individual teachers will need to find their own methods, and use them judiciously (one should beware of using methods simply because one likes them). It is important also to find specific ways of increasing motivation, i.e. make the course relevant and interesting to the particular students you have. That is why it is a good idea to try to arrange that the classes are fairly uniform and to get to know your students.

In conclusion, it is one of the theses of this book that attention to some quite prosaic details will enable better teaching to be achieved.

Finally I feel that some way of recording the successes and failures of courses should be standardised in each department, so that we can learn from our mistakes. It is this topic that will be addressed in the next chapter.

13. Assessment of Teaching

Preview

Some assessment of teaching is desirable. The problem is that assessment is one of the most difficult tasks. To make things worse, it is a task which usually receives scant attention. This chapter seeks first to give a simple way of recording the results of each course, which are then available in subsequent years to give some guidance. Secondly it aims at briefly discussing how a new method of teaching should be assessed. The most obvious pitfalls are sketched.

13.1 The need for external control

Any assessment of teaching requires questionnaires. It is to be hoped that each lecturer has sufficiently close contact with students to be able to notice when a course is going disastrously and when a complete change in teaching method is essential. It certainly makes sense to ask individual students how they react after the first few lectures, and yet, even with asking, thoroughly experienced teachers can be off the mark. Consequently a questionnaire is helpful.

The diagnostic test already mentioned in §8.2 should help avoid a complete misunderstanding of the level of competency of the students, to help them brush up their skills, and help to ensure that the level of the course matches the students' knowledge, i.e. that the first rule of teaching is satisfied. But nevertheless, it is important to have a questionnaire early on in the course, say after a week or so, to check that the course is being well understood.

It is best to have a questionnaire which can be answered quickly. The questionnaire should contain questions which will enable you to make changes, i.e. the questions asked should preferably be very practical. An example of such a questionnaire can be found in Appendix D.

It is quite common in Sweden to have an elaborate questionnaire at the end of the course, which often takes the student an half an hour or more to complete. To get as many people as possible to complete it, one uses some of the lecture period. I feel this is a waste of teaching time, and it is better to use a shorter and simpler questionnaire.

It is important that the control of the questionnaire is not left solely to the lecturer involved. The lecturer should seek assistance from a colleague or a student, somebody who can independently look at the replies to the questionnaires and make sure the lecturer deceives neither himself nor anybody else, and also somebody who can discuss how to interpret the results.

Lecturers have far too few controls on their work. In many departments, the only control is if the students come in a group to complain, by which time it is much too late to do something. Of course, the tradition is that lecturers are their own severest critics, and this is often the case, but there are always some black sheep.

It should be the duty of a senior member of staff to come to at least part of a lecture given by each member of staff, perhaps once a year. Of course such a member of staff has to have the right personality. He should not come to comment in general, but rather to detect if the lecturer is right off the mark.

If the practice of having a mentor (§9.2) is in operation, then this process can begin with new lecturers, most of whom will welcome some feedback and re-assurance that their courses are going satisfactorily.

I can remember what happened in my very first lecture course, which was to a group of second year students. The students were deeply unhappy, and came to complain to the head of department, who asked to see the notes they had taken during my lectures. This was a course where no set book was prescribed, and the students were expected to learn mainly from the lectures. Finding that my notes were perfectly satisfactory, clear, well laid out and at the right level, the head of department judged the complaints to be frivolous, and did not inform me. I was too imperceptive to observe that the students were unhappy. If I had carried out at least two of the recommendations I have made above, this state of affairs would not have arisen. The first was a diagnostic test. This would have revealed the fact that I had been unaware of, that the students were for the most part extremely weak. And of course, the quick questionnaire would have revealed the difficulties. But even before these had occurred, the mentor would have taken me aside, knowing that these second year students had only passed the first year by the slimmest of margins. Instead I spent a lot of time preparing courses which failed to satisfy the first rule of teaching, and which left the students deeply unhappy. I would have also have realised the dangers if the records of the previous year had been given me. Which is the point I wish to discuss now, how to keep records in a sensible way.

13.2 Records

It is sensible and useful to compile records in a consistent way.[13] I propose doing what companies do at the end of the year, they put their results down in columns, with the previous year's results on the left-hand column, and the present year's results on the right-hand column (see Fig. 13.1). In order to get a consistent approach it is necessary to decide on certain standard presentations.

For instance, it would be sensible to have an estimate of the difficulty of the examination, and so one could have a difficulty index, which could be chosen to be 100 in a given fixed year. This remains the reference from one year to another.

It would be worth while adding certain remarks, such as the approach used, and what one thought the results indicated. Of course, all of these remarks are not in any sense objective. Still, they give some guidance, and enable one to benefit from one's own experience and the experience of others.

Note that in Fig. 13.1 both years have their own columns. It is thus possible to compare one year with the other, taking account of the work done both by students and the lecturer, and also taking account of the difficulty of the examination.

Of course most of this description is subjective, and it should not be used in any formal way, certainly not by the head of department nor by the administration. There is no way in which a reasonably fair and objective assessment can be made. Such assessments are not the point of this exercise. The method outlined here aims at providing an informal way of recording and subsequently discussing the quality of the results achieved, so that there is the possibility of finding improvements.

---

[13] One lecturer said that he found the practice of writing a report after each lecture a useful discipline.

Course: Calculus 2	**1994**	**1995**
Calculus 1 results, number of passes.	26	30
Number of students attending class/lectures at beginning of course	55	56
Number of students attending class/lectures at end of course	35	38
Overall difficulty of course (1990 = 100)	90	85
Lecturer's work load (hours)	200	250
Student's work load (hours)	170	180
Students' questionnaire rating of course (max 6)	3	3.5
Diagnostic test % pass rate	25	30
Degree of directed teaching (§8.5)	3	3
Final examination % pass rate	55	69
Difficulty of examination (1990 = 100)	85	80
Attainment levels (maximum = 6) (§2.5)		
Abstraction level	3	3
Definition level	3	3
Techniques level	4	4
Proof level	3	3
Ingenuity level	3	3
Rules of teaching (maximum = 6) (§7.2)		
Rule 1 Right level	3	3
Rule 2 Uniform class	3	3
Rule 3 Deal with the obvious	4	4
Rule 4 Active students	3	3
Rule 5 Make demands	3	2
Rule 6 Students must be able to ask questions	2	3
Rule 7 Motivate	2	4
Rule 8 Lectures must be interesting	3	4
Rule 9 Students are people	3	3
Rule 10 Lecturers must keep on learning	4	4
Totals (Rules of teaching & attainment levels)	46	49

Fig. 13.1 Calculus 2

Notes

(1) Book used: Adams, "Calculus", Third edition. (used in both 1994 and 1995).

(2) Comments 1994: A somewhat lack-lustre course, with bored students.

(3) Comments 1995: This year I added some motivation by every so often indicating where the course may be of use in the students' main subject, electrical engineering, and even quoted Maxwell's equations. This motivation was useful, did not take too much time, but still needs improvement.

(4) Solid improvements of the examination results were achieved at the end of the course with a summary based mainly on doing very simple versions of the standard questions, in which problems were chosen so that calculations and manipulations were as simple as possible, thus putting the emphasis on the ideas rather than on technical skill.

(5) The examination was also notably easier than the previous year and this helped the results. Nevertheless the questions that appeared on the examination paper were sufficiently varied and difficult for somebody who passed well to illustrate mastery of the subject. Also the range of marks was quite large.

(6) Again there was a large drop off in the number who began the course and the number who continued till the end. As yet I have not explanation for this. When I first noticed this drop off, the students who were present suggested it was due to some students staying at home to watch the results of the Olympic games on TV. The following week the explanation was that a number of students had influenza. This requires careful investigation next year.

13.3 Assessing new methods of teaching

The above methods are meant to cover courses in which there is no radical change in the method of instruction, but it is also important to compare different methods of teaching, and this importance will increase as new techniques of teaching, especially those involving computers, the internet and multi-media are introduced.

We shall be concerned with the most obvious points, and what I have written here should be regarded as a beginning, not as a complete answer. Evaluating is in itself an extremely difficult task. It requires a very rare sort of person, namely one able to avoid prejudice, one who has good judgement, and it takes a lot of time and expertise to build up the competence to make a good evaluation. A number of fundamental indices need to be checked, and an awareness of what can go wrong is required. Perhaps in time a reliable, rigorous standard approach to evaluation will emerge.

New methods of teaching are being introduced all the time. The evaluation of such new methods needs to be of an extremely high standard, and it is not sensible to rely, as we mainly seem to do, on an evaluation carried out by the innovator of the new method of teaching.

In medicine there is a fairly well agreed method of evaluation. Thus there is always a control group, and medicines are administered by a double-blind system, where neither the doctor administering the medicine nor the patient knows whether the tablet is a placebo or the medicine being tested.

In agriculture too it is well accepted that one must be careful about testing crops, that one designs the plots so that unsuspected advantages or disadvantages do not influence the result. So it is common to use magic square arrangements.

In teaching however there do not seem to be any universally accepted methods. Teaching is not easier to assess, it is harder. It is harder because it involves human beings. For instance, a poorer course might give just as good results as a better course, but only because students work very much harder. However, if students work much harder in one course, they may end up failing another course.

13.4 Assessing as it is done now

In practice, there are only a few properly conducted experiments on education. There are a large number of experiments carried out by individual lecturers. A typical such experiment would go as follows. The innovator of the new method puts it into action, then gives the students a questionnaire and comes to a conclusion. This cannot be accepted as a reasonable method of assessment, there being a natural bias in favour of any new ideas one introduces oneself.

I think of assessing as a combination of the methods used by auditors to assess accounts and the methods a consumer organisation might use to compare say various refrigerators.

Auditors always like to see the accounts organised into various categories, and compare the results of one year with another. They then check any figures which seem to be out of line, and expect a good explanation for any discrepancy. They sum horizontally and vertically, and check that the sums agree. They are aware that mistakes can easily occur, as well as fraud.

Similarly an assessor should be on the look-out for errors and fraud. It must be remembered that the innovator of a method of teaching is desperately keen to get it approved, and can unconsciously or deliberately cheat.

A consumer organisation such as "Which" in England and "Rån och Rön" in Sweden would normally make a list of the important characteristics of a given product, say refrigerators, and then arrange to test the refrigerators using the same criteria. A careful analysis of the advantages and disadvantages then follows. Often there is no over all winner: refrigerator A consumes less electricity than refrigerator B, which however tends to have a longer life span, or less harmful effects on the environment, for instance. Such a consumer organisation will of course listen to the comments of the manufacturer of each fridge, but will in addition seek independent means of checking.

I therefore make the following minimum suggestions:

(1) The assessment should be thought out before the experiment is carried out.

(2) The person in charge of assessing should be independent.

(3) There must be some method by which the assessment can be independently checked.

(4) As many things as possible should be kept constant, so that one knows which cause is having which effect.

(5) A control group is essential.

(6) Careful measurements of the fundamental indices (see §13.5 below) must be carried out before and after the experiment begins.

13.5 Fundamental indices

What are the fundamental indices, the items we need to know about and which need to be carefully measured at every new investigation into the success of teaching? There is going to be disagreement but I suggest the following:

(1) The knowledge and ability of the students at the beginning of the course.
(2) The time taken for the students to study the course.
(3) The time taken for the lecturer to prepare and deliver the course.
(4) The material learnt.
(5) The ability to apply the ideas to unfamiliar situations.
(6) How many students obtain a really high standard.
(7) How many students achieve a satisfactory standard.
(8) Student assessment of the course.
(9) History of the course, i.e. what happened in previous years.

Much of what is required could be put in the form described in §13.2.

13.6 Balance

Having obtained these fundamental indices for the experimental and the control group it is now time to assess the advantages and disadvantages. This is among the most difficult part, and of course is subjective, for it depends on the weight attached to each of the desirable effects. For example, if system A produces 10% of really good students and in total 60% satisfactory students, is it better or worse than system B with 5% good students but in total 90% satisfactory students?

Most systems will have some advantages and some disadvantages, and a careful analysis of these will be required.

Learning to assess takes time and effort, and if there were a national group of assessors, the whole process would be more efficient and of course more uniform. So I would think it sensible for a country to appoint a group of say ten assessors who could carefully assess the large number of different trials of new methods taking place. Such assessors should of course take time to canvass the opinion of the other staff at the institution where the new system has been introduced. It is also important to remember that a questionnaire can lead to misleading results, and so the students should be consulted independently.

One must not underestimate the time and effort required for a proper evaluation. A minimum requirement is to have an independent assessor responsible from the start for the evaluation, a control group, and careful collection of data. Most changes of system have some advantage, the question is whether they are on the whole worthwhile. Good judgement, unbiased consideration, and most of all, clarity, is what is required.

13.7 Rules when assessing a new method of teaching

My above remarks can be conveniently be summarised in the following rules:

The first rule is that one should not trust a method in which the person who has introduced the new method is the evaluator. It is not reasonable to expect that somebody who has spent many months working on a new project can dispassionately evaluate it. It is not human nature.

My second rule is don't trust an experiment without a control group. That provides the only chance of comparing two different methods of teaching. A control group is essential. We use a control group in medicine, why not in teaching?

The third rule is that the experiment should measure all the relevant factors. In particular, the work loads of both students and staff must be considered.

The fourth rule is that when assessing a teaching experiment make sure you include a history of the subject. The same trial may have already been done in the past, and it makes sense to take advantage of the experience of others.

The fifth and final rule is to ask the staff and the students what they think of the experiment, even those not involved. If something untoward has been done, some rumour of it will often appear in the common room. For instance, it is not unknown for the person in charge of the project to give some or even all of the exam paper away before the examination[14], and one can get to know of it by chatting to some of the staff or students.

---

[14] This does lead to very much better results than could normally be achieved, although some say cynically that it makes no difference.

Let us apply these rules to three experimental methods of teaching, stretching back in time from the present to a quarter century ago, when I began a two year experiment.

In brief, in the experiment I introduced, the idea was that students should read a book on their own, with the help of notes and advice, and be examined by the teacher, i.e. a modified Keller approach.

Then I had no control group. I, the person who had devised the experiment, was the one who evaluated it. My experiment did not pass the minimum of the rules. Nor did it pass the third rule, namely that the experiment should measure all the relevant factors. It did not for instance assess how much work each lecturer did, nor how much work each student did, with reference to the standard method of teaching then in vogue.

So in listing the rules, I am quite aware that I have been guilty of not observing them myself.

Then I go to an experiment reported by Hubbard [1990]. The starting point of this experiment is the criticisms of the lecture as a teaching system by Gibbs. Hubbard therefore suggests her own method, which again is basically that students read the book themselves, and get help and be examined. There is some group work and tutorial work.

Finally I refer you to a project at Mälardalens Högskola, Larfeldt [1998]. In essence this too is a system without lectures, complete with problems to be handed in.

Three experiments, spanning just over 25 years, all with much the same idea, i.e. eliminate lectures and see whether learning improves. And all three suffer from having unconvincing assessments, all failing on the basic three rules, and also on the fourth rule: when assessing a teaching experiment make sure you include a history of the subject.

13.8 Comparing the quantity of material in courses

I was once involved in assessing a method of teaching linear algebra, method A, which had replaced method B, and which it was claimed was far superior. Indeed, the examination pass rate with method A was 80% while that for method B the year before was 70%, a clear improvement. The teaching expert who was attached to the course did not understand the mathematics in the course. I wanted to find a way of convincing her that method A contained very much less material than method B, it being my contention that a simpler and shorter course was the real explanation for the improved pass rate.

As both of the courses were based on text-books, an unbiased method was to count the number of exercises, examples, theorems, definitions and pages. Of course if one uses these figures in a mechanical way, one will come to something unreliable. One may need to put in various factors, for instance if certain theorems in one book are considerably more difficult than others. When the courses one is comparing involve the same subject matter, like the courses I was investigating, this method is very helpful. Comparing the two courses in Linear Algebra I obtained the following figures:

	Method A	**Method B**
Exercises	178	365
Examples	58	231
Theorems	39	90
Definitions	19	62
Pages	189	286

Fig. 13.2 Counting material

These figures convinced me that Course B contained double the material of Course A.

The subject matter in Course A was all included in Course B, and the various items in Course B should really have been weighted more, as they were harder, more abstract, and more sophisticated. But I used the same weighting.

When I carried out this counting I was helped by somebody who had been involved in teaching both courses. (Doing this type of evaluation on your own is not a good idea, and you should also avoid getting a like-minded individual to help.) I myself had taught only course B, but had carefully studied the course notes and descriptions for course A. The two of us had no difficulty in coming to agreement about the totals. I myself liked this method of comparing the two courses because it did not rely solely on subjective assessments.

The attainment levels described in §2.5 can be judged and used to estimate a course. In the case of the two courses described above:

Level	Abstraction	Definition	Techniques	Proof	Ingenuity
Method A	2	3	4	4	2
Method B	3	4	4	5	3

Fig. 13.3 Attainment levels in Courses A and B

13.9 Assessing the difficulty of examinations

In comparing examinations in the same subject I have found it useful to use the following techniques. I count the number of theorems and definitions each question in the examination paper involves. Each question is also evaluated as easy, standard, or harder. These steps are carried out firstly for all the problems in the examination paper, and then for the minimum number of problems which are needed for a pass. When taking the minimum number for a pass, one chooses the easiest problems.

In the case of Courses A and B above I came to the following conclusions:

	Number required		Classification of questions		
	Theorems	Definitions	Easy	Standard	Harder
Course A, all questions	5	19	4.8	3.2	0
Course B, all questions	9	10	0	8	0
Course A pass only	3	8	3.2	0	0
Course B pass only	4	3	0	3.6	0

Fig. 13.4 Comparison of two examinations

(For comparability, some mark normalisation was required. This explains the decimals.)

Classifying questions as standard is relatively easy. These are the questions which are covered and explained in the course, and which appear frequently in past exam papers, and which we have noticed from experience that students find relatively easy. If they are unusual questions, or if they involve difficulties which have not been covered before, then they may legitimately be classified as harder problems instead. Easy questions are ones which are obviously easier than the standard questions, for example questions which basically require one step.

For instance, for linear algebra, to diagonalise a symmetric matrix would be regarded as standard, unless there were difficulties say in calculating eigenvalues which required a certain amount of ingenuity. To orthogonalise a set of vectors using the Gram-Schmidt method would be regarded as standard. However, to find the eigenvalues and eigenvectors of a 2x2 symmetric matrix would be described as easy, whereas a 3x3 matrix would probably constitute a standard question..

The totals in Fig. 13.4 reveal a clear difference between the difficulty of the examinations.

One can also assess the examination by using the list in §8.3, and also the attainment levels in §2.5.

13.10 Southampton University's approach

Southampton university had an interesting and enterprising approach to evaluating their own teaching performance.

They invited a mathematician from another university to spend three or so days looking at and assessing their department and then discussing his assessment with the staff. This exercise was repeated three or four years later.

They then established reviews every five years, with a specially appointed group, consisting of two members of the department and one member who had some interest in mathematics from another department. The group would attend some lectures, interview students and made suggestions on such issues as the content of the courses and how they linked up with other courses, and the types of activities and learning situations to which students were exposed.

13.11 Summary

It is customary when introducing a new system to put most of the effort into introducing the new system, and very little into evaluating it. But a proper evaluation is very important, for if we do not know when we have an improvement and why it is an improvement we have no rational reason for making a change. Even modest improvements of say 1% could save enormous sums of money. The trouble is that new methods of teaching seldom show a clear-cut advantage. A new car carburettor may show a consistent 1% improvement, but teaching is not like a car carburettor.

Since few experimental methods involve sufficiently careful evaluation by independent people of good judgement, you should be cautious when examining these new methods.

Despite the difficulty of making thorough and valid assessments, it is still worthwhile making some record of the results in your courses from year to year in a standardised way, for comparison purposes. It is not conclusive, but better than nothing at all.

13.12 Epilogue

Once when I was a young man I met George Polya (whose books and ideas of teaching I have strongly recommended) and told him that I had tried his methods with only limited success. He smiled gently, and said, "We do the best we can." And that, dear reader, is all we can do. The rest, and it is the most significant part, depends on the determination, the interest, the ability, the intelligence and the hard work of the student. There never was and there still is no royal road to mathematics. I hope nevertheless that this book will be of some help.

APPPENDICES

Appendix A	Education Systems in Brief
Appendix B	Videos
Appendix C	Quotations
Appendix D	Quick questionnaire
Appendix E	Alternative courses
Appendix F	Cartoons
Appendix G	Maxims
Appendix H	Projects

Appendix A Education Systems in Brief

A.1 The English education system

Fig. A1.1 below summarises the position of most people in England at various ages:

Age	Type	Qualifications at end
5-11	Primary school	None
11-16	Secondary school	GCSE
16-18	Sixth form	A or A/S level
18-21	University or other Higher Education	B.A. or B.Sc.

Fig. A1.1 Stages in the English education system

Education till age 16 is compulsory, although many (37% in 1996) now stay on till 18.

GCSE (General Certificate of Secondary Education) is taken at 16 or so. After this students will typically take three A-levels, or two A-levels and two AS-levels. The AS level is approximately half the content of the A-level, and it was introduced in 1988, to overcome the specialised education which results from so few subjects being studied. A broader five A-level pattern remains under discussion.

There is a discontinuity between GCSE and A-levels which causes much hardship. Basically what happened was that the GCSE replaced O-levels, but the A-levels were left unchanged.

The student who is majoring in mathematics will spend about 4 or 4½ hours of mathematics lessons per week in their final year at school.

Admission to higher education is competitive and mainly based on A-level performance.

Higher education usually consists of the following degrees:

Bachelor's degree: a 3-year course. The final degree is usually classified as either a First, Upper Second, Lower Second, or Third. Most students then stop their higher education at this stage. Those that continue their studies proceed as follows:

Master's degree: A one-year course with possibly a short dissertation. A few students will then go on to the degree of Doctor of Philosophy, which is at least three years of research. Students who show promise in the beginning of the Master's degree can skip the Master's degree and proceed directly with the Doctorate.

Among the 700,000 18-year olds, 20% enter higher education, 3,000 take mathematics, perhaps combined with another subject. That is about ½% of the age group, and approximately 5% of those taking mathematics at A-levels. (Porter [1992].)

Mathematics courses will be taught by academic mathematicians whose main interest is mathematics itself. Engineering students will also usually be taught mathematics by academic mathematicians, although these courses are concerned with mathematics that the aspiring engineers will find useful in their discipline. There is some pressure to reduce the mathematical topics covered and make them more applicable to the particular type of engineering being studied.

A serious problem of English education highlighted by Porter [1992] is the large number of people who stop studying mathematics at age 16 and find they do not have the background to understand either the world they live in or else the disciplines they subsequently study.

Students are financed by a mixture of loans and grants, and help from their parents. The grant part has been steadily reduced. Compared to those in employment, students are poor, and have difficulty in making ends meet. Postgraduates in particular are badly off.

A.2 The Swedish education system in brief

The most striking part about the Swedish education system is that it starts so late, in spite of very early child-care. There is a tendency for mothers to go back to part-time or full-time work when their child is a year old, which is possible because of an excellent subsidised child-care system.

Age	Type	Qualifications at end
1½—7	Dagis (child-minding)	None
7—16	Grundskola	Grundskola
16—19	Gymnasium	Student examen
19—25	University or other higher education	Diploma, Bachelor's degree, Master's degree

Fig. A2.1 Stages in Swedish education system

There is some schooling during this child-care, but it is not regarded as an important part of formal education. Genuine school is delayed till the age of 7. Thus Swedish children start their schooling quite a lot later than English children. School attendance till 16 is compulsory, but 95% stay on voluntarily till the age of 19 (source Swedish Institute fact sheet March 1995).

Unlike in England, students in their final three years study a large number of subjects. The total hours of instruction is given as 2150 (Skollagen, Bilaga 2, SFS 1991:1107). Of this mathematics was given as 300 hours, about 14%. There is about 9% of study time which can be chosen freely. Languages are also an important part of school studies, with Swedish about 9%, English 7%, and a second language at about 9%. Physics is about 10%, Chemistry about 8% and Biology 5% of the time. There is a bit of Philosophy. Thus a hodgepodge of subjects (nearly 17), with very little specialisation.

So those students interested in science, engineering, or mathematics will spend some 60 hours on mathematics lessons in the year before university, some 2 hours per week.

Admission to higher education for those under 25 years of age is based on the school performance, or alternatively, the passing of an entrance examination. To be admissible to higher education courses requires that the applicant has passed at least 90% of the required school courses. Previously most Swedish men were expected to have military training, which delayed their university studies, but now this training is no longer available for all .

For those 25 years and over there are very few academic requirements (see §4.1).

Something like 65,000 students enter higher education each year; by comparison the number of 20-year-olds is of the order of 114,000.

The duration and extent of university programmes are expressed in a system of points, one point representing a week of full time study. Thus one academic year of successful full-time study gives 40 points.

The first type of degree is the diploma (högskolexamen), requiring a minimum of 80 points. The second type is the bachelor degree (kandidatexamen), obtained after completion of 120-140 points of study, with a minimum of 60 points in the major subject. The master's degree (magisterexamen) requires 160 points including 80 points in the major subject and a single thesis of 20 points or two theses each of 10 points.

Postgraduate degrees include the Licentiate which requires a further two years after completing the first degree, and the Doctor's degree which requires a further four years after the first degree. The additional title of docent can be obtained for further significant research achievements.

In Sweden, enormous numbers of students study mathematical courses, but these are mainly engineering students. Something like 10-15% of the university students study some mathematics at their university, indeed, it is among the most studied of subjects. By contrast, there are essentially no courses for genuine mathematicians.

Students are financed by student loans, which are fairly generous. Students although not rich, certainly are not poor, and live reasonably comfortable lives. The loans are not given in subsequent years unless the student has managed to pass sufficiently many courses (which tends to concentrate the mind), and loans must be re-paid, a process which can take much of a working life.

Doctoral students are expected to teach as well. Partly as a result of the amount of teaching and partly through tradition, doctorates are obtained quite late in life. The mean age for obtaining a doctorate in mathematics is 33; in the humanities and social sciences the mean age is over 40.

A.3 The German education system

This continues to be a high-status and elitist system of education.

Admission to the university requires passing the Abitur, a local school examination with wide variations in content and level. It is not uncommon for school children with poor results to repeat a year. Admission to university is determined not by each university but by a central body, and not all candidates can be given places, so some may have to apply repeatedly. Consequently the age of entry usually lies between 20 and 25. Thus the ages in the table reflect the minimum rather than the practice.

Age	Type	Qualification
6—10	Grundschule	
11—19	Gymnasium(academic) OR Real Gymnasium (classical academic) OR Gesamtschule (comprehensive) OR Technical gymnasium	
19—22/23	University OR Fachhochschule	Staatsexam (teaching diploma) OR Diplom

Fig. A3.1 German education system

The university, which stresses theoretical studies, has been supplemented by an alternative, called the Fachhochschule, which stresses engineering and science and is meant to be more immediately applicable. In both institutions teaching and research are regarded as indivisible.

Diplom is about the standard of an English masters and usually takes 4 years, and often requires a thesis. The Ph.D. usually takes a further 3 years full time.

Financial support is by student loans and a grant (now being reduced). The Ph.D. is often supported by a teaching or an assistant position. Fees at the universities are negligible.

After the doctor's degree comes a further qualification, the Habilitation. This usually takes about 6 years and involves a substantial amount of research. Only after one has completed the habilitation has one the right to become a lecturer, although with a Ph.D. alone one can get a job as assistant lecturer, and this can be permanent.

Typically in the first year at university one studies demanding courses which take a full academic year, courses such as Linear Algebra, or Analysis. There are usually four or five such courses. The Staatsexam, which is for those wanting to become teachers, has the same status as the Diplom, and is not easier. German students will in addition to attending courses take part in seminars. Here each student must prepare a seminar, typically based on a research paper, e.g. a paper on group rings, or something from an advanced research book. The student is expected to attend and participate in a number of seminars. One is a pro-seminar, i.e. a simpler piece of mathematics which the student must lecture on. Attendance and contributions at the seminars given by the other students is required as well.

Students register for their examinations independently of the courses. There are minimum course requirements, but no set periods by which student must be examined. This tends to extend study times.

A.4 The French system in brief.

There are few nations which take education quite so seriously. Education plays an essential part throughout one's life. Compulsory schooling lasts from 6 or 7 till 15.

The final qualification at school is the baccalauréat, of which there are a large number of choices, the most prestigious of which is the bac C, the mathematics and physical science option. The baccalauréat is a state examination with identical papers throughout France, and in general is more taxing than the English A-level. Currently the baccalauréat is being watered down and some prestigious schools are transferring to the international baccalauréat.

Age	Type	Qualification at end
6/7—10/11	Primary	
11/12—14/15	Collège	
15/16—17/18	Lycée	Baccalauréat
18—20	University	DEUG 2years
18—22		Maîtrise 4 years
18—26		DEA 8 years (doctorate)
18—20 (varies)	Grande École préparation 1 or 2 years minimum	
20—25	Grande École	

Fig. A4.1 French education system

Admission to the university is automatic with the baccalauréat. Admission to a Grande École, which are the most prestigious of the education institutions, is by competitive national examination. Students may prepare for the Grande École at university.

Failure rates in examination are high (50%) so that most students take longer to complete their courses than Fig. A4.1 suggests.

Appendix B Videos

B.1 Videos

Videos are available from the many mathematical associations. As examples I list some videos from the AMS, LMS, and MAA. Very often these lectures are more difficult to follow than ordinary lectures, but many of them are very inspiring.

B.2 From the AMS

1) Qubba for Al-Kashi—Yvonne Dold-Samplonius
2) Preparing for Careers in Mathematics—Annalisa Crannell, Steven J. Altschuler, William Browning, Ray E. Collings, Margaret Holen, Sandra L. Rhoades, James R. Schatz, Anita Solow, DePauw, and Francis Edward Su.
3) Fermat's Last Theorem—Barry Mazur
4) Celebrating 100 Years of Meetings: History and Reminiscences
5) The Seiberg-Witten Invariants
6) The New Shepherd's Lamp—Jean-Pierre Bourguignon
7) In Search of Symmetry—William Browder
8) The Current Interface of Geometry and Elementary Particle Physics—I. M. Singer
9) Optimization of Extended Surfaces for Heat Transfer—J. Ernest Wilkins
10) Fermat's Last Theorem—The Theorem and its Proof: An Exploration of Issues and Ideas—Will Hearst, Robert Osserman, Lenore Blum, Karl Rubin, Kenneth Ribet, and John H. Conway
11) Special Package Offer! Fermat's Last Theorem—The Theorem and its Proof: An Exploration of Issues and Ideas and Modular Elliptic Forms and Fermat's Last Theorem

12) Modular Elliptic Curves and Fermat's Last Theorem—Kenneth A. Ribet
13) Characteristic Forms—Shiing S. Chern
14) Interview with I. M. Gelfand
15) Some Mathematics of Baseball—Henry O. Pollak
16) Wavelets Making Waves in Mathematics and Engineering—Ingrid Daubechies
17) A New Look at Knot Polynomials—Joan Birman
18) The Problem of Scale in Ecology—Simon Levin
19) Coloring Knots—Sylvain Cappell
20) Combinatorial Reconstruction Theorems—Ronald L. Graham
21) Cosets, Clusters, Spinsters, and the Schröder-Bernstein Theorem—Paul Halmos
22) Descartes and Problem Solving—Judith Grabiner
23) Fiftieth Anniversary Meeting of the Metropolitan New York Section of the Mathematical Association of America
24) Pedagogical Peeves and Other Complaints of Age: Crazy Al, Still Teaching Calculus after All These Years—Al Novikoff
25) The Teaching of Calculus: Careful Changes—Gilbert Strang
26) Computational Crystal Growers Workshop—Jean E. Taylor
27) Algorithms in Algebraic Number Theory—H. W. Lenstra, Jr.
28) Compound Soap Bubbles, Shortest Networks, and Minimal Surfaces—Frank Morgan
29) Laplacians of Graphs and Hypergraphs—Fan R. K. Chung
30) Mathematics under Hardship Conditions in the Third World—Neal I. Koblitz
31) On the Maximum Principle—Louis Nirenberg
32) Physics and the Mysteries of Space—Michael Atiyah
33) The Theory and Applications of Harmonic Mappings between Riemannian Manifolds—Richard M. Schoen
34) Lambda-Trees and Their Applications—John W. Morgan
35) A Century of Representation Theory of Finite Groups—Charles W. Curtis
36) Addresses on the Works of the 1990 Fields Medalists and Nevanlinna Prize Winner—Michio Jimbó, Joan S. Birman, Ludwig D. Faddeev, and Lázló Lovász
37) Algebra as a Means of Understanding Mathematics—Saunders Mac Lane
38) Applications of Non-Linear Analysis in Topology—Karen K. Uhlenbeck

39) Birational Classification of Algebraic Threefolds—Shigefumi Mori
40) Braids, Galois Groups and Some Arithmetic Functions—Yasutaka Ihara
41) Case Studies of Political Opinions Passed Off as Science and Mathematics—Serge Lang
42) Computational Complexity of Higher Type Functions—Stephen A. Cook
43) Computing Optimal Geometries—Jean E. Taylor
44) Dynamical and Ergodic Properties of Subgroup Actions on Homogeneous Spaces with Applications to Number Theory—Grigorii Margulis
45) Elliptic Methods in Variational Problems—Andreas Floer
46) Fifty Years of Mathematical Reviews—Saunders Mac Lane
47) From Coxeter Diagrams to Kummer Identities—George Daniel Mostow
48) Gauge Theories and the Jones Polynomial—Edward Witten
49) Geometric Algorithms and Algorithmic Geometry—László Lovász
50) Harish-Chandra and His Work—Rebecca Herb
51) Hyperbolic Billiards—Yakov G. Sinai
52) Intersection Cohomology Methods in Representation Theory—George Lusztig
53) Multidimensional Hypergeometric Functions and Their Appearance in Conformal Field Theory, Algebraic K-Theory, Algebraic Geometry, etc. Alexandre N. Varchenko
54) Natural Minimal Surfaces Via Theory and Computation—David Hoffmann
55) Pseudodifferential Operators, Corners and Singular Limits—Richard B Melrose
56) Recent Work on Motifs—Spencer Bloch
57) The Art of Renaissance Science: Galileo and Perspective—Joseph W. Dauben
58) The Conformal Field Theory from the View of the Cohomology Theory of the Lie Algebras—Boris L. Feigin
59) The Interaction of Nonlinear Analysis and Modern Applied Mathematics—Andrew J. Majda
60) Videotapes from ICM 90
61) Viscosity Solutions of Partial Differential Equations—Michael Crandall
62) Von Neumann Algebras in Mathematics and Physics—Vaughan F. R. Jones
63) $ax^2 + hxy + cy^2 = n$—John H. Conway

64) Applications of PDE Methods by Gromov, Floer, and Others to Symplectic Geometry—Dusa McDuff
65) Crystals, in Equilibrium and Otherwise—Jean E. Taylor
66) Fifty Years of Eigenvalue Perturbation Theory—Barry Simon
67) Nonwellfounded Sets and Their Applications—Jon Barwise
68) Some Applications of Group Representations—Nolan Wallach
69) The Transition to Chaos: The Orbit Diagram and the Mandelbrot Set
70) Arithmetic Progressions: From Hilbert to Shelah—Ronald L. Graham
71) Chaos, Fractals and Dynamics: Computer Experiments in Mathematics
72) Georg Cantor: The Battle for Transfinite Set Theory—Joseph W. Dauben
73) Indeterminate Forms Revisited—Ralph P. Boas
74) Introducing Mathematica—Stephen Wolfram
75) Mathematical Problems of Liquid Crystals—Haim Brezis
76) Some Major Research Departments of Mathematics—Saunders Mac Lane
77) The Beauty and Complexity of the Mandelbrot Set—John Hubbard
78) The Flowering of Applied Mathematics in America—Peter D. Lax
79) The Story of the Higher-Dimensional Poincaré Conjecture (What Actually Happened on the Beaches of Rio de Janeiro)—Stephen Smale
80) The Topological Constraints on Analysis—Raoul H. Bott
81) Transonic Flow and Mixed Equations—Cathleen S. Morawetz
82) European Mathematicians' Migration to America—Lipman Bers
83) How Computers Have Changed the Way I Teach—John G. Kemeny
84) Oscar Zariski and His Work—David Mumford
85) Addresses on the Work of the 1986 Fields Medalists and Nevanlinna Prize Winner—Michael F. Atiyah, Barry Mazur, John W. Milnor, and Volker Strassen
86) Classifying General Classes—Saharon Shelah
87) Complexity Aspects of Numerical Analysis—Stephen Smale
88) Efficient Algorithms in Number Theory—Hendrik W. Lenstra
89) Geometry of Four-Manifolds—Simon K. Donaldson
90) New Developments in the Theory of Geometric Partial Differential Equations—Richard M. Schoen
91) Problems in Harmonic Analysis Related to Oscillatory Integrals and Curvature—Elias M. Stein
92) Quasiformal Mappings—Frederick W. Gehring
93) Recent Progress in Arithmetic Algebraic Geometry—Gerd Faltings

94) Representations of Reductive Lie Groups—David A. Vogan, Jr.
95) Soft and Hard Symplectic Geometry—Mikhael Gromov
96) String Theory and Geometry—Edward Witten
97) Underlying Concepts in the Proof of the Bieberbach Conjecture—Louis de Branges
98) Episodes in the Origins of the Representation Theory—Thomas Hawkins
99) Matrices I Have Met—Paul Halmos

B.3 From the MAA

1) Let us Teach Guessing, George Polya
2) John von Neuman, a Biography
3) Courant in Göttingen and New York
4) The Moore Method, a documentary on R. L. Moore
5) Pits, Peaks and Passes, Part 1, Marston Morse
6) Pits, Peaks and Passes, Part 2, Marston Morse
7) MAA Calculus Films in Video Format, 3 tape collection
8) The Theorem of Pythagoras
9) The story of Pi
10) Similarity
11) Sines and Cosines, Part 1
12) Sines and Cosines, Part 2
13) Sines and Cosines, Part 3
14) Polynomials
15) The Teachers Workshop
16) The tunnel of Samos
17) Was Newton's Calculus just a dead end? Maclaurin and the Scottish Connection, Judith V Grabiner
18) The Seventy-fifth Anniversary Celebration
19) Mathematics and Computation: Proliferation and Fragmentation
20) Has progress in mathematics slowed down? Paul Halmos
21) The last 75 years: Giants of Applied Mathematics. Cathleen S. Morawets
22) Problems for all seasons, Ivan Niven
23) The contributions of mathematics to education. Peter J Hilton
24) An analogue of Huber's formula for Riemann's Zeta function, Floyd L Williams

25) Developing the next generaton of mathematicians, Uri Tresiman

B.4 From the LMS

1) Wallpaper patterns in different geometries, A F Beardon
2) Chaology, M Berry
3) How mathematics gets into knots, R Brown
4) Geometry and Computers, P Giblin
5) Games that solve problems, W A Hodges
6) How should a mathematician think about shape? D G Kendall
7) The rise and fall of matrices, W Ledermann
8) Games animals play, J Maynard Smith
9) Codes and Cyphers, F C Piper
10) Stamping through mathematics, R J Wilson
11) Geometry and Perspective, E C Zeeman

Appendix C Quotations

C.1 Quotations

The occasional use of a quotation can be a pleasing way to add interest to a lecture. (I stress occasional. One can easily become a bore.) Quotations can be obtained by looking at text books which treat the same subjects you are treating, from history of mathematics books and from:

(1) "Comic sections : the book of mathematical jokes, humour, wit,and wisdom," by Desmond MacHale, Boole Press
(2) "Out of the Mouths of Mathematicians," Rosemary Schmalz, published by the MAA. It is a collection of over 700 quotes grouped into 16 topics, one of which is humor.
(3) "Memorabilia Mathematica by Robert Moritz," first written in 1914, re-published by the MAA.
(4) "Calculus Gems," Simmons G F, 1992, McGraw Hill
(5) The internet, e.g.
`http://math.furman.edu/~mwoodard/mquot.html`

One way of collecting quotations is to keep an eye open all the time and make a note whenever one reads something which seems suitable for quoting.

In the mean while, here are a few quotations to give the flavour.

Algebra
"Important though the general concepts and propositions may be with which the modern and industrious passion for axiomatizing and generalizing has presented us, in algebra perhaps more than anywhere else, nevertheless I am convinced that the special problems in all their complexity constitute the

stock and core of mathematics, and that to master their difficulties requires on the whole the harder labor." Herman Weyl.

"Modem algebra has exposed for the first time the full variety and richness of possible mathematical systems." Birkhoff G and MacLane S.

Analysis

"If a nonnegative quantity was so small that it is smaller than any given one, then it certainly could not be anything but zero. To those who ask what the infinitely small quantity in mathematics is, we answer that it is actually zero. Hence there are not so many mysteries hidden in this concept as they are usually believed to be. These supposed mysteries have rendered the calculus of the infinitely small quite suspect to many people. Those doubts that remain we shall thoroughly remove in the following pages, where we shall explain this calculus." Leonhard Euler (1 707-1783)

"There is in mathematics hardly a single infinite series of which the sum is determined in a rigorous way." N H Abel

Applications

"You never know when some abstract mathematical theory, the formulation and solution of which you have long since forgotten, can not only prove to be of value, but even the decisive solution to the very problems you are struggling with today." Anon.

"Mathematical theories from the happy hunting grounds of pure mathematicians are found suitable to describe the airflow produced by aircraft with such excellent accuracy that they can be applied direcly to airplane design." Theodore von Karman

(Describing the Universe) ".. that cannot be understood unless one first learns to comprehend the language and read the letters in which it is composed. It is written in the language of mathematics." Galileo.

Calculus

"But just as much as it is easy to find the differential of a given quantity, so it is difficult to find the integral of a given differential. Moreover, sometimes we cannot say with certainty whether the integral of a given quantity can be found or not." Bernoulli, Johann

"The Mean Value Theorem is the midwife of calculus -- not very important or glamorous by itself, but often helping to deliver other theorems that are of major significance." Purcell, E and Varberg, D.

Complex Variable

"The power and significance of Cauchy's theorem—the centre-piece of complex analysis—is, I believe, best revealed initially through its applications." Priestly H A.

"Often problems that do not appear to involve complex numbers are nevertheless solved most elegantly by viewing them through complex spectacles." Needham T.

Differential Equations

"Historically, differential equations arose from humanity's interest in and curiosity about the nature of the world in which we live." Giordano F R & Weir M D

"The phenomenon of resonance was also responsible for the collapse of the Broughton suspension bridge near Manchester, England, in 1831. This occurred when a column of soldiers marched in cadence over the bridge thereby seting up a periodic force of rather large amplitude." Braun, M.

Examinations

"Examiners' inspiration is students' despiration." Ted Hurley

"This year's examination was much harder than last year's." Most students, most years.

General

"Mathematics possesses not only truth, but supreme beauty—a beauty cold and austere, like that of sculpture, without appeal to any part of our weaker nature... sublimely pure, and capable of a stern perfection such as only the greatest art can show." Bertrand Russell.

"Reasoning about infinity is one of the characteristic features of mathematics as well as its main source of conflict." John Stillwell

Geometry

"At the age of eleven, I began Euclid... This was one of the great events of my life, as dazzling as first love." Bertrand Russell

"Inspiration is needed in geometry, just as much as in poetry." Pushkin, A S

Group Theory

"You are very lucky today. I am going to tell you not one joke but two. Two isomorphic jokes." Stewart Stonehewer

"Galois introduced into the theory the exceedingly important idea of a self-conjugate sub-group, and the corresponding division of groups into simple and composite. Moreover, by shewing that to every equation of finite degree there corresponds a group of finite order on which all the properties of the equation depend, Galois indicated how far reaching the applications of the theory might be, and thereby contributed greatly, if indirectly, to its subsequent developement." Burnside W.

Linear Algebra

"We do not allow matrices to multiply unless they have been properly introduced," when remarking the number of columns of A should be equal to the number of rows of B before defining the product AB. Perlis S

"In the present century linear algebra has acquired new richness and versatility through the use of the concepts of group and non-commutative ring in algebra itself, and through the use of infinite-dimensional function spaces in analysis. Applications to quantum mechanics stimulated a still more rapid development of the theory of these spaces, which has become one of the most important parts of contemporary functional analysis." Mal'cev A I.

Learning Mathematics

"We do our students a disservice if we ask little of them." Professor Richard Askey, University of Wisconsin-Madison

"Don't just read it; fight it! Ask your own questions, look for your own examples, discover your own proofs. Is the hypothesis necessary? Is the converse true? What happens in the classical special case? What about the degenerate cases? Where does the proof use the hypothesis?" Halmos, Paul R.

Numerical Methods

"The purpose of computing is insight, not numbers." Richard Hamming,

Recursion

They found a skeleton in a narrow cavity wall in the middle of an old house being pulled down in Dublin. Round its neck was a medallion bearing the inscription “Irish hide-and seek champion 1927.” Desmond McHale, radio talk.

"To iterate is human; to recurse, divine."

"To define recursion, we must first define recursion."

Relativity

“The phrase was still in vogue that ‘only 3 people understand Relativity’ at a time when Eddington was complaining that the trouble about Relativity as an examination subject in Part III was that it was such a soft option.” J E Littlewood.

Research

“Mathematics is the transformation of coffee into theorems.” Alfred Renyi

“Research is what I do when I don’t know what I am doing.”

Statistics

“Lottery, n.: A tax on people who are bad at mathematics.”

(Thanks to the following for suggesting quotations: R Booth, T Hawkes, T Hurley, G Keady, W Potter, R Schmalz, P Schulz, T Ward.)

Appendix D Quick Questionnaire

D.1 Quick questionnaire

COURSE EVALUATION
Give values from 1 to 5 to the following questions.

A) Generally
1 = The lectures are working badly. 5 = The lectures are working very well.

B) Difficult/easy
1 = The course is too easy. 5 = The course is too hard.

C) Problems
1 = The problems are too easy. 5 = The problems are too difficult.

D) Audibility
1 = The lectures are difficult to hear. 5 = The lectures are easy to hear.

E) Visibility
1= The writing on the white-board is difficult to read. 5 = The writing on the white-board is easy to read

Appendix E Alternative Courses

E.1 Alternative courses

A summary of innovations in mathematics courses in the United Kingdom was prepared in 1993 for the London Mathematical Society by Tim Porter of the University College of North Wales, Bangor. This has been revised subsequently, and the full report is available from the London Mathematical Society.

The innovations ranged from new methods of teaching traditional mathematical subjects, new topics, teaching mathematics history, and teaching "soft mathematics" in the sense which I have defined it in §5.2. A few examples are given below:

(a) Ideas of mathematics (Tim Porter, Bangor)

Arranged in 5 blocks:

(1) Symbols and Algebraic manipulation
(2) Limitations of mathematics, logic, etc.
(3) Algorithms
(4) Modelling Mechanics
(5) Problem solving

The idea is to discuss rather than to lecture, in an attempt to improve the intuition of students, their presentation skills, and problem formulation and problem solving skills. Assessment is by a combination of essays, examinations, and 'Lab' reports.

(b) Mathematical Modelling (D Thatcher, de Montfort University, Leicester)

A year long course with lectures and tutorials on formulating and analysing mathematical models of situations (models mainly from outside classical applied mathematical areas).

(c) Mathematical Problem Solving (Don Collins, Q.M.W., University of London)

A third year course. A list of problems is provided which the students must solve. The written solutions are assessed and there is an additional oral examination. The course is one eighth of the year's work. About 40 students take the course, which is taught by two staff members.

(d) Mathematics in Society (Ian Porteous, Liverpool University)

A third year course consisting of ten class meetings each of 2 hours, led by different speakers. Students also prepare projects under supervision. These projects are presented both orally and in writing. Topics include "medical health statistics", and "do females underachieve in mathematics."

Appendix F Cartoons

F.1 Cartoons

Just like a quotation, a cartoon can enliven a lecture. But cartoons should be used even more sparingly than quotations. I employ at most one cartoon per lecture course. This is partly because I cannot draw cartoons very well, but also because they can become tedious.

By keeping a sharp eye out, you can gradually over the years collect cartoons. You can also draw them yourself. If you can't draw very well, you can easily learn to draw the occasional cartoon by using the methods suggested by Nelms [1964] of "creative tracing." I also suggest the very small book Bradley [1990], which is meant for children, but is helpful. As an example, when I introduce the concept of a division ring, I say it is the marriage of a multiplicative group and an additive group, together with their two little children, the associative laws, and draw Fig. F1.1.

Fig.F1.1 A division ring

Appendix G Maxims

Instead of repeating the same thing over and over again, in other words nagging, I use maxims. You may say that is still nagging, which is true, but it is not so annoying.

G.1 Time, organisation and study

(1) Do not say "I'll study when I have the time", since you may never have the time.
(2) Time is like water in a sponge, the harder you squeeze the more you get.
(3) Summarise, let nothing evade your eyes. I *think it is important to get students to summarise their subject matter, and so I use this plagiarism from Tom Lehrer.*
(4) Learn a theorem each night and a definition before light.
(5) A revision in time is worth nine
(6) Active repetition is the backbone of knowledge.
(7) Prepare before each class, work hard after.

G.2 What is important

(1) Thou shallt honour your definitions and theorems for the rest of your days.
(2) A theorem in the head is worth two in the book.
(3) A definition without an example is like a fiddle without a string.
(4) A fool and his theorems are soon parted. (i.e. you must know your theorems and have them at hand when you need them.)

G.3 How to study

(1) The tools for reading a book are pencil and paper.

(2) A person who asks a question is a fool for 5 minutes; a person who never asks a question is a fool for the rest of his life.

(3) I hear and I forget
I see and I remember
I do and I know

(4) The best way to learn is to teach.

G.4 Solving problems

(1) Always start with the unknown and the given.

(2) Move the beginning to the end and the end towards the beginning. i.e. reason forward from the conditions and backwards from what you are trying to prove, to see if you can get the conditions and conclusions closer to one another.

(3) All is fair in love, war and mathematics, i.e. use your initiative and not necessarily the set methods you have learnt.

(4) If you can draw a picture you may get somewhere.

(5) Haven't I seen you somewhere before? This is a comment useful at a party to try to meet somebody you fancy. But it can be used effectively to find out what is relevant to the problem in hand.

(6) Do the mathematical two-step, guess- check; guess-check.

(7) A definition is not only the meaning of a word, it is also a tool to solve problems.

(8) Once you have your answer two things remain: find a way of checking it and find a way of expressing it.

G.5 General advice to lecturers

(1) The shy do not learn, and the strict can not teach.

(2) One can't proceed from the informal to the formal by formal means.

(3) If you think you are arguing with a fool—the chances are he thinks the same.

(4) Do not give the answer to a problem nobody has asked.

(5) "The most important single factor influencing learning is what the learner already knows. Ascertain this, and teach him accordingly." David P Ausebel.

I thank James Davenport , Avinoam Mann, Robert F Morse, and Trevor Hawkes for helping compile these maxims.

Appendix H Projects

Here are some of the projects which Southampton University Mathematics Department offered to students in the academic year 1999/2000. Although only a fraction of their projects, this list indicates how varied projects can be. These projects obviously take considerable student and staff time, but are exciting and very valuable. (Thanks to Gareth Jones for sending this material to me.)

Drainage of iron ore stockpiles

Iron ore mining companies in Australia are interested in determining how much water is held in their stockpiles. After the ore has been cut from the ground it is processed, and water is added to allow crushing to occur and to prevent large amounts of dust being created. The ore is then left in very large piles (20 meters high and half a kilometer long). It is important for them to know how much of the water will drain out and if they should take any precautions in making the pile in the first place to aid the drainage. The project will involve the student in understanding and modelling the movement of water in a granular material. Literature on groundwater movement will need to be reviewed. The model will probably take the form of partial differential equations and these may be studied by either analytical methods or numerical solution. This project might be of particular interest to a student interested in working on problems of direct interest to industrial companies. It requires a mix of problem formulation, analytical methods and numerical solutions. The projects are sufficiently large that the interested student can choose to balance the emphasis of their effort into any of these three parts of the work but will require some effort in all parts.

Population genetics of human evolution

Tracing human origins is a multidisciplinary task involving archaeology, anthropology, genetics and sophisticated mathematics. Modern molecular genetics enables the differences between human subpopulations and individuals to be quantified and reassembled into plausible maps of ancient populations and population migrations. In this project you will use computer simulation methods to examine data from mitochondrial DNA (transmitted only maternally) to reconstruct the human family tree.

Dynamical systems and cosmology

Even if one makes assumptions about the symmetries of our universe it can be very difficult to find exact solutions of Einstein's equations with those symmetries. However in the study of certain types of cosmology it is possible to reduce the Einstein's equations to a non-linear system of ordinary differential equations. Although it is usually still not possible to solve the resulting equations exactly, one can use techniques from the study of dynamical systems to understand the qualitative behaviour of the solutions. This project will look at how one can make such a reduction to a finite-dimensional dynamical system and use these methods to study models of both stable and chaotic cosmologies.

Knot polynomials

A knot is a continuous injective map from a circle to three-dimensional space. Two knots are equivalent if the image of one map can be moved continuously in R^3 to take up the position of the image of the second map. The classification of knots (i.e. describing a procedure to determine if two knots are equivalent) is one of the most interesting and challenging unsolved problems in topology. A number of polynomial invariants of knots have been discovered, several in the last ten years. The project is to describe some of these polynomials and use them to distinguish between various knots.

The universal graph

A graph G with countably many vertices is said to be universal if, whenever Γ and Δ are disjoint finite sets of vertices of G, there is a vertex in G which is adjacent to all the vertices in Γ, and to none of those in Δ. This apparently simple property has the surprising consequence that any two universal graphs are isomorphic, so there is essentially just one universal graph G. On the other hand, Erdös and Rényi showed that almost all countable graphs (in a sense which can be made precise) are universal, and hence they are isomorphic to G! The aim of the project is to find simple proofs of these facts, to find natural examples of universal graphs, and to investigate their symmetry properties.

Dirichlet series and primes in an arithmetic progression

One of the most important results in number theory is Dirichlet's theorem that an arithmetic progression $\{a + nd: a, d > 0, n = 0,1,2,\ldots\}$ with a and d coprime contains an infinite number of distinct prime numbers. One of the reasons why this theorem is so celebrated is that it was one of the first important theorems in number theory that relied on analytic tools in its proof.

These analytic tools are concerned with a family of infinite series called Dirichlet series which define analytic functions of the form

$$f(s) = \sum_{n=1}^{\infty} a_n n^{-s}$$

where s is a complex number. (For example if $a_n = 1$ for all n then we get the famous Riemann zeta function.) In this project students will be expected to explain enough of the theory of Dirichlet series so that they can prove Dirichlet's theorem. An interesting topic they will probably cover is the theory of Euler products. For a longer project covering two semesters they might look at modular forms which played such a crucial role in the proof of Fermat's Last Theorem.

Bibliography

Allenby [1983]; "Rings, Fields and Groups", Allenby R B J T, Edward Arnold.

Andersson [1996]; "Two is one too many: Dyadic memory collaboration effects on encoding and retrieval of episodes". Andersson J, Linköpings Studies in Education and Psychology No. 47

Anderson [1989]; "Introduction to flight", Anderson J.D. (Jr.) , McGraw Hill

Barry [1990]; "A Core Curriculum in Mathematics for the European Engineer", M D J Barry and N C Steele, SEFI, The European Society for Engineering Education.

Beard [1984]; "Teaching and Learning in Higher Education", Beard R M., Hartley J, Harper and Row

Bell [1973]; "Present and future in higher education", Bell & Youngson

Bonner [1995]; "Distributed Multimedia University: From Vision to Reality", Bonner R, Berry A, Marjanovic O., Ascilite '95 pp 36-43.

Bradley [1990]; "How to draw cartoons", Bradley S, Archer R, Henderson Publishing

Brittanica [1987]; "Encyclopedia Brittanica", Micropaedia 12, page 165.

Burton [1995]; "Technology in Mathematics Teaching", Burton L and Jaworski B (editors), Chartwell-Bratt

Burn [1992]; "Numbers and Functions: Steps to Analysis", Burn R P, Cambridge University Press.

Burn [1997]; "Teaching Undergraduate Mathematics", Burn B, Appleby J, Maher P (editors), Imperial College Press

Buzan [1989]; "Use your head", Buzan T, BBC Books.

Clark [1960]; "The Art of Lecturing", Clark G K, Clark B, Heffer.

Cobal [1993]; "The University as an Institution today", Cobal A B

Courant [1996]; "What is Mathematics", Courant R, Robbins H, revised by Stewart I, Oxford University Press.

Davis [1983]; "The Mathematical Experience", Davis P J, Hersh R, Penguin Books.

Dawkins [1996]; "Climbing Mount Improbable", Dawkins R, W.W. Norton & Co.

Dubinsky [1992]; "The concept of Function: Aspects of Epistemology and Pedagogy", Dubinski E & Harel G (eds), MAA Monographs.

Dunham [1994]; "The Mathematical Universe", Dunham W, John Wiley and Sons

Earle [1989]; "Graphics for engineers", Earle, J. H., Addison-Wesley Publishing Company

Eble [1976]; "The craft of teaching", Eble, K.E. ,Jussey-Bass

Engineering Council [1985]; "The changing mathematical background of undergraduate engineers", R Sutherland and S Pozzi, Engineering Council, London

Ernest [1991]; "The Philosophy of Mathematics Education", Ernest P, The Falmer Press

Freudenthal [1978]; "Weeding and Sowing: preface to a science of mathematics education", Freudenthal H, Reidel Dordrecht.

Gibbs [1982]; "Twenty Terrible Reasons for Lecturing", Gibbs G, SCEDSIP, Occasional Papers 8.

Giancoli [1998]; "Physics principles with applications", Giancoli D, C, Prentice Hall International, Inc.

Glatthorn [1993]; "Learning Twice", Glatthorn A, Harper Collins College Publishers

Griffiths [1974]; "Mathematics Curriculum and Society", Griffiths and Howson, A.G., Cambridge University Press.

Gullberg [1997]; "Mathematics: From the Birth of Numbers", Gullberg, J, W.W. Norton and Company

Haggarty [1993]; "Fundamentals of Mathematical Analysis", Haggarty, R, Addison-Wesley

Halsey [1995]; "Change in British Society: from 1900 to the present day", Halsey A H, Oxford University Press.

Herstein [1996]; "Abstract Algebra", Herstein I N, Prentice Hall International.

Higginson [1988]; "Advancing A-levels", Higginson G, HMSO

Hoffman [1965]; "Linear Algebra", Hoffman K and Kunze R, Prentice-Hall

Howson[1998] ; "The value of comparative studies", A. G Howson, in "International Comparisons in Mathematics Education" edited by Kaiser, G, Luna, E. and Huntley, I, Falmer Press, 1998, 165-188.

Hubbard [1990]; "Tertiary mathematics without lectures", Hubbard R, International Journal of Mathematical Education in Science and Technology, 21(4), pp 567-571

Hubbard [1991]; "53 Interesting Ways to Teach Mathematics", Hubbard R, Technical and Educational Services Ltd.

Huff [1993]; "How to Lie with Statistics", Huff, Darell, W.W. Norton & Company

IMA [1995]; "Mathematics matters in engineering", Institute of Mathematics and its Applications, IMA, Southend.

I Physics [1994]; "The case for change in post-16 physics: planning for the 21st Century", Institute of Physics 16-19 Working Party.

James [1993]; "Representations and Characters of Groups", James G and Liebeck M, Cambridge University Press.

Kaplansky [1969]; "Infinite Abelian Groups", Kaplansky I, Ann Arbor

Katz [1993]; "A history of mathematics", Katz V J, Harper Collins

Kemeny [1964]; "Random essays on mathematics, education and computers", Kemeny J G, Prentice Hall

Kline [1972]; "Mathematical Thought from Ancient to Modern Times", Kline M, Oxford University Press.

Kordemsky [1972]; "The Moscow Puzzles", Kordemsky B A, Penguin Books, 1972
Krantz [1999]; "How to teach mathematics", (Second Edition), Krantz, S.G., American Mathematical Society
Lange [1993]; "Innovation in maths education by modelling and applications", Lange J, Huntley I, Keitel C, Niss M, Ellis Harwood
Larfeldt [1998]; "Matematikprojektet vid Mälardalens Högskola", Larfeldt T, Axner U
Laurilard[1993]; "Rethinking University Teaching", Laurilard D, Routledge.
LMS [1992]; "The Future for Honours Degree Courses in Mathematics and Statistics", Committee under the chairmanship of Peter Neumann. Available from the London Mathematical Society
LMS [1995]; "Tackling the Mathematics Problem", Committee under the chairmanship of Howson. A G. Available from the London Mathematical Society
Mason [1982]; "Thinking Mathematically", Mason J, Burton L & Stacey K, Addison Wesley.
Mayhew [1990]; "The Quest for Quality", Mayhew L B., Ford P J, Hubbard D L, Jossey-Bass
McKeachie [1978]; "Teaching tips", McKeachie, W J., Heath
McLone [1973]; "The Training of Mathematicians", R R McLone, Social Science Research Council report
Michener [1978]; "Understanding Understanding Mathematics", Michener E, Cognitive Science 2, pp361-383.
Miller [1970]; "Success, failure and wastage in higher education", Miller, S W,
Nelms [1964]; "Thinking with a pencil", Nelms H, Barnes and Noble.
Norman [1995]; "Introduction to Linear Algebra", Norman D, Addison.Wesley Publishers Limited
Northedge [1997]; "The Sciences Good Study Guide", Northedge A, Thomas J, Lane A, & Peasgood A, The Open University.
Paulos [1991]; "Beyond Numeracy", Paulos J A
Paulos [1995]; " A mathematician reads the newspaper", Paulos J. A., Basic Books.
Parsons [1976]; "How to Study Effectively", Parsons C, Arrow Books Limited
Phillips [1991]; "Perspectives on Learning", Phillips D C. & Soltis J F., Teachers College Press 1991.
Polya [1965]; "Mathematical Discovery", Polya G, John Wylie and sons
Polya [1977]; "Mathematical methods in Science", Polya G
Polya [1985]; "How to solve it", Polya G, Princeton University Press
Quinney [1996]; "Calculus Connections", Quinney D, Harding R. John Wiley and Sons, 1996.
Ramsden [1992]; "Learning how to teach in higher education", Ramsden P, Routledge
Sawyer [1943]; "Mathematician's delight", Sawyer W W, Penguin
Schoenfeld [1985]; "Mathematical Problem Solving", Schoenfeild A H, Academic Press

Schoenfeld [1989]; “Exploration of Students’ Mathematical Beliefs and Behaviours”, Schoenfeild A H, Journal of research in Mathematics Educaton, 20 (4) pp338-355
Schoenfeld [1990]; “A source book for college mathematics teaching”, Schoenfeld A H (editor), The Mathematical Association of America.
Sierpinska [1994]; “Understanding in Mathematics”, Sierpinska A, Falmer Press
SNCRC [1995]; “International Review of Swedish Research in Mathematical Science 1995”, International Review Committee, Swedish Natural Science Research Council
Stewart [1975]; “Concepts of Modern Mathematics”, Stewart I, Penguin Books
Stewart [1987]; “The Problems of Mathematics”, Stewart I, Oxford University Press.
Stewart [1998]; “Counting the Pyramid Builders”, Stewart I, p 76, Scientific American Sept 1998
Stillwell [1989]; “Mathematics and its History”, Stillwell J, Springer.
Stillwell [1992]; “Geometry of Surfaces”, Stillwell J, Springer.
Tall [1991]; “Advanced mathematical thinking”, Tall D O (editor), Kluwer Academic Publishers
Turnbull [1962]; “The Great Mathematicians”, Turnbull H W, Methuen
Watson [1998]; “Questions and Prompts for Mathematical Thinking”, Watson A & Mason J, ATM
Wigner [1960]; “The unreasonable Effectiveness of Mathematics in the Natural Sciences”, Wigner E P,Communications in Pure and Applied Mathematics. 13,1-14 (1960).
Wirsma [1995]; “Research methods in education”, Wirsma W, Allyn & Bacon 1995.

Index

A

abstraction level · 17
academic year organisation · 98
academic, life of · 40
aims · 23
alchemists · 49
algorithm · 55
alternative methods of teaching · 59, 127
American education system · 4
asking questions · 69, 149
assess students · 99
assessment of teaching · 183
attainment levels · 17
axiomatic courses · 118

B

baccalauréat · 204
back-substitution · 179
Badura · 82
Baer · 169
barrage of questions method · 170
Behaviourism · 81
blackboard · 126
Bok, Derik, quotation from · 27
Bonner · 61
books for teaching ideas · 139
bread · 178
breaks in lectures · 160
Burn · 139

C

calculators · 62
calculus course · 144
checklist of school mathematics · 7
Chinese proverb · 82
City Explorer · 70
clever students · 30
compactification · 156
computers · 24, 62, 124
concrete, computer makes more · 124
concrete, making argument more · 159
confusion of words · 176
conic sections · 160
consistent set of headings · 166
construction lines · 143
constructivism · 81
content of mathematics courses · 52
continuous assessment · 103
continuous function · 176
control group · 191
Cornell method for notes · 69
Crooke's rays · 27

D

definition level · 17
definitions · 76, 154
degree of generality · 159
demand · 88
democratisation, of education · 16
department leadership · 114
departmental organisation · 111
dependence table · 69
detective story · 169
determination · 19
development from axioms · 119
Dewey's theories · 81
diagnostic tests · 103
diagonalisation · 169
directed teaching · 105
discover-it-yourself · 83
discussing theorems · 168

E

Education Acts · 11
education system in the USA · 4
educationalists · 23
eigenvector with eigenvalue · 152
eighth fundamental rule · 90
elite · 30
employment · 23
encouragement · 90
enthusiastic · 141
entrance examination · 201
equivalence relations · 172
essential example · 156
essential subjects for undergaduates · 57
Euclid's algorithm · 125
evaluation · 188
examinations · 12, 38, 98
expected value definition · 174
external controls · 184

F

Fachhochschule · 203
Fermi problems · 53
fifth fundamental rule · 88
first fundamental rule · 83
flying blind · 120
formal · 181
formulae lists · 62
fourth fundamental rule of teaching · 86
French system · 204
functions · 172
fundamental indices · 190
fundamental rules of teaching · 83

G

general students · 52
German · 203
Gestalt · 81, 148
Giancoli · 123
good picture · 155
government requirements · 37
Griffiths · 140
group teaching · 127
guess-check · 180
guessing · 73
guided reading · 130

H

Halsey · 13
hard mathematics · 52
Higginson · 14
higher education, current definition · 13
How to Solve It · 75
Hubbard · 192

I

indirect methods of teaching · 134
informal definition · 154
ingenuity level · 17, 18
introduce ideas · 152
intuitive idea · 155

J

Jig-saw Puzzle Method · 70

K

Kaplansky · 159
Keller method · 129
Kemeny · 16, 50, 85

L

large classes · 48
lecture-based · 121
lecturer · 40
lecturer's general approach · 139
lecturer's manual · 114
lectures · 86
library · 134
Lighthill · 33
limit. · 174
linear independence · 176
lists of formulae · 62
Littlewood · 171
Locke's blank tablet theory · 81

M

Mårtenson · 55
mass education · 12
mathematical history · 56
mathematical two-step · 74
mathematics specialists · 55

maxims · 159
McLone · 28, 57, 88
memorise · 68
mentors for students · 113
method of assessment · 188
methods courses · 118
mnemonics · 68
modern university · 12
motivate · 150

N

neat handwriting · 166
neglected teaching aid · 104
Nevile · 61
new concepts · 152
ninth fundamental rule · 90
Northedge · 140
notation · 171

O

Open University · 61
oral examinations · 101
organisation in education · 97
organise a break · 160
outstanding graduates · 21
overhead projector · 126

P

Piaget's theories · 82
Plato's recollection theory · 81
point-set topology · 147
Polya · 74
Porter · 52
prep · 162
problem based learning · 132
problems class · 163
programmed learning · 127, 132
programmed learning type books · 122
project · 130
projective planes · 120
proof · 19
proof level · 17, 18

Q

questionnaire · 184
Quinney · 61

R

Ramsden · 15, 106
read a book · 69
records · 185
reducing the quantity · 48
repetition · 158
re-sits · 99
retiring member · 116
revision · 79

S

Sawyer · 15
Schaum · 121
Schoenfeld · 139
second fundamental rule · 84
seminar course · 133
seminars, research · 91
sixth fundamental rule of teaching · 88
smart students · 20
Socratic discussion · 83
soft mathematics · 52
solving problems · 73
Southampton University · 195
spike · 178
spoon.feeding · 105
spread-sheet · 125
$SQERPSR^2$ · 70
staff/student committee · 112
standards · 17, 19
student attainments · 20
student mentor · 113
student/staff ratio · 16
student's most common fault · 75
student-centred · 82
students' manual · 114
study skills · 67
stupid students · 34
styles of explaining ideas · 160
super book, computer use · 125
Swedish government · 37, 39

T

teacher wrong · 181
teacher-centred · 49
teaching improvement · 48

teaching in groups · 129
techniques level · 17
technological methods of teaching · 59
tell the whole story · 169
tenth fundamental rule of teaching · 91
text-based · 121
text-books · 121
theories of learning · 81
third fundamental rule of teaching · 85
torch · 160
total load · 116
training and education · 31
tricky topics · 172
tutorials · 161
two definitions, formal and informal · 76

U

unemployment · 24
university town · 26

V

voice · 165
Vygotsky · 82

W

what does the examination test? · 104
Wigner · 33